CW00950517

THE FACE OF THE EARTH
The Legacy of Eduard Suess

Judd, J. W.: Prof. Eduard Suess, For. Mem. R.S.,
Nature, 93, 1914, p. 246

Suess held much the same position among German-speaking peoples as did Huxley among English and Americans. They both held that, in addition to their scientific labours, however exacting these might be, something in the way of service was due to the cities in which they lived and the states to which they belonged. In 1862 Suess had directed attention to the unsatisfactory condition of the water-supply of Vienna, and, from 1863 to 1873, he was called upon to serve as a member of the Municipal Council of Vienna; it was due to his initiative in this capacity that an aqueduct, 110 kilometres long, was built to bring water from the Alps to the city, and that other great improvements in the sanitary conditions of Vienna were undertaken.

Schuchert, Ch.: Eduard Suess,
Science, 39, 1914, p. 934

The greater part of Suess's long life was devoted to working out the evolution of the features of the earth's surface. The problem of mountain-building presented itself to his mind during his many excursions in the eastern Alps, and in 1875 he stated his views thereon in the small volume called DIE ENTSTEHUNG DER ALPEN, an octavo of 168 pages. Up to this time his publications numbered sixty titles, his studies having ranged over nearly all the branches of geology.

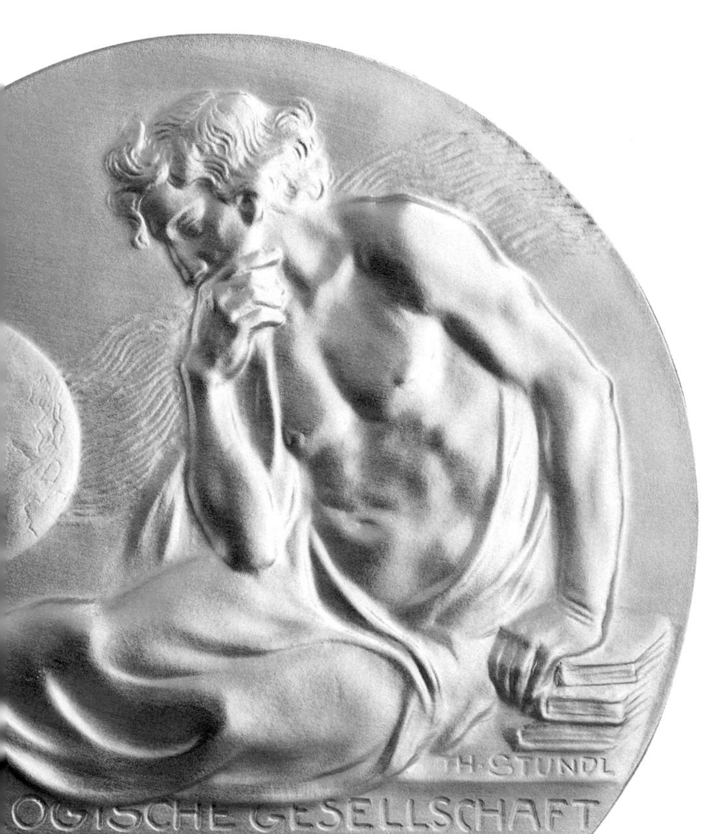

Eduard Suess Medal

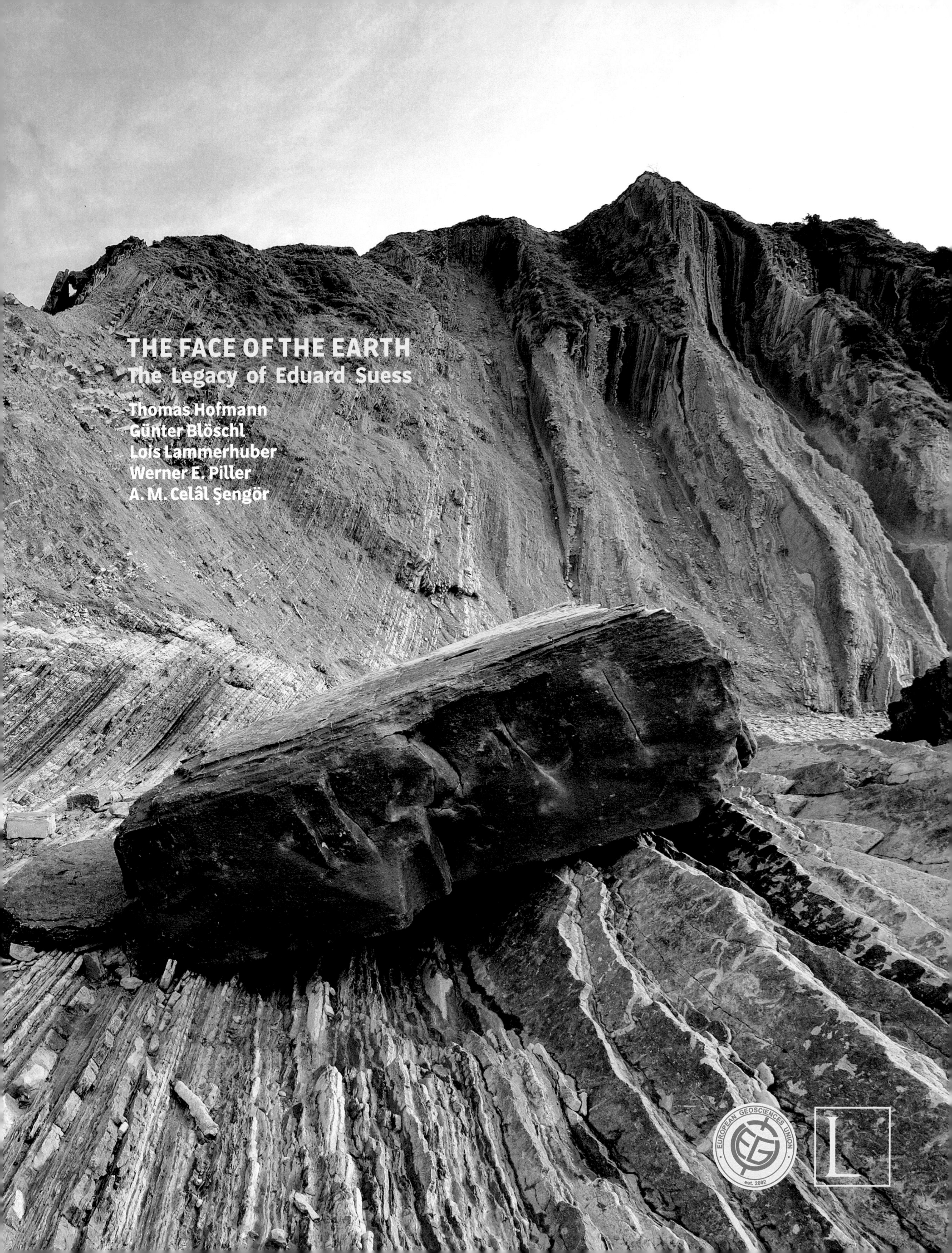

THE FACE OF THE EARTH
The Legacy of Eduard Suess

Thomas Hofmann
Günter Blöschl
Lois Lammerhuber
Werner E. Piller
A. M. Celâl Şengör

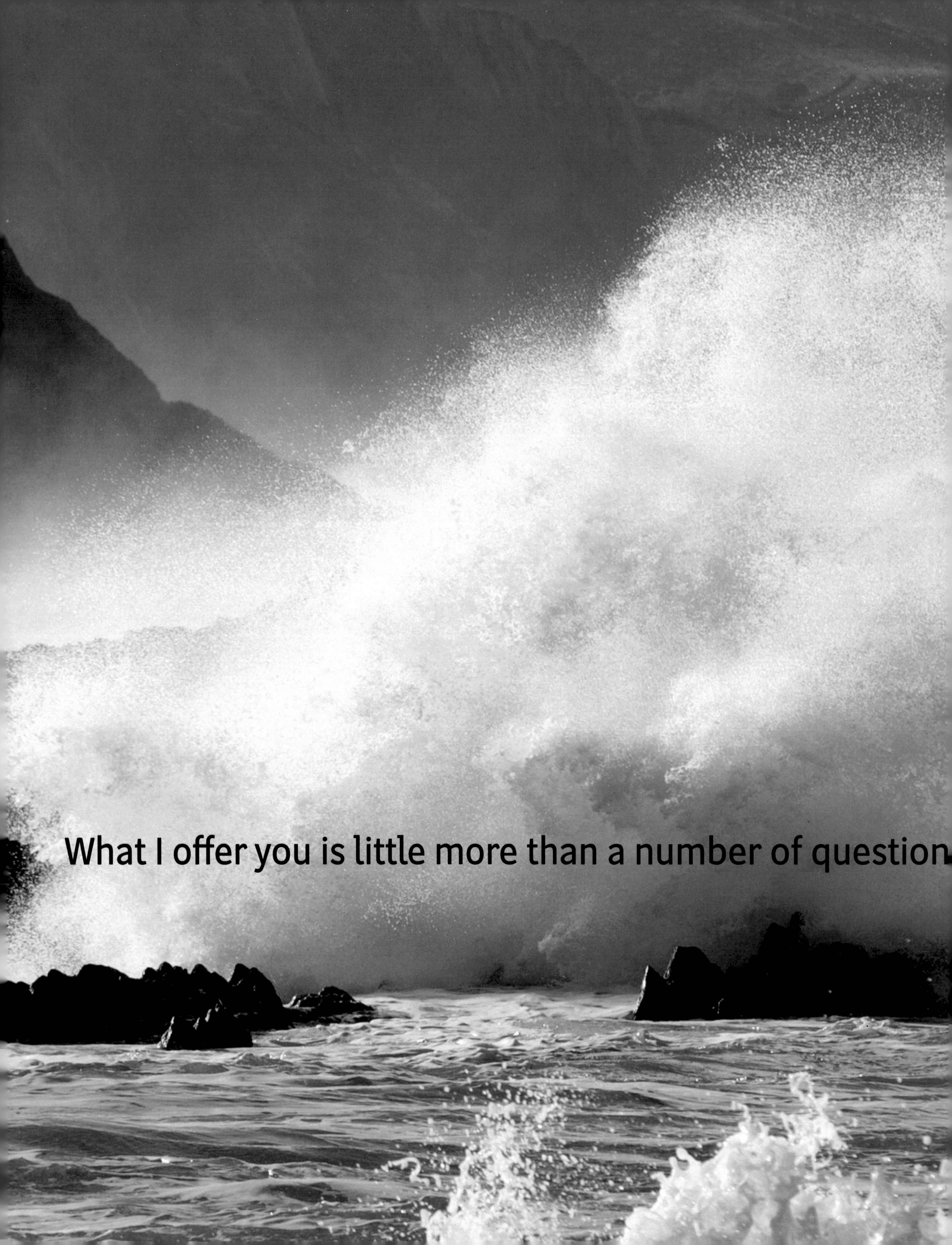

What I offer you is little more than a number of question

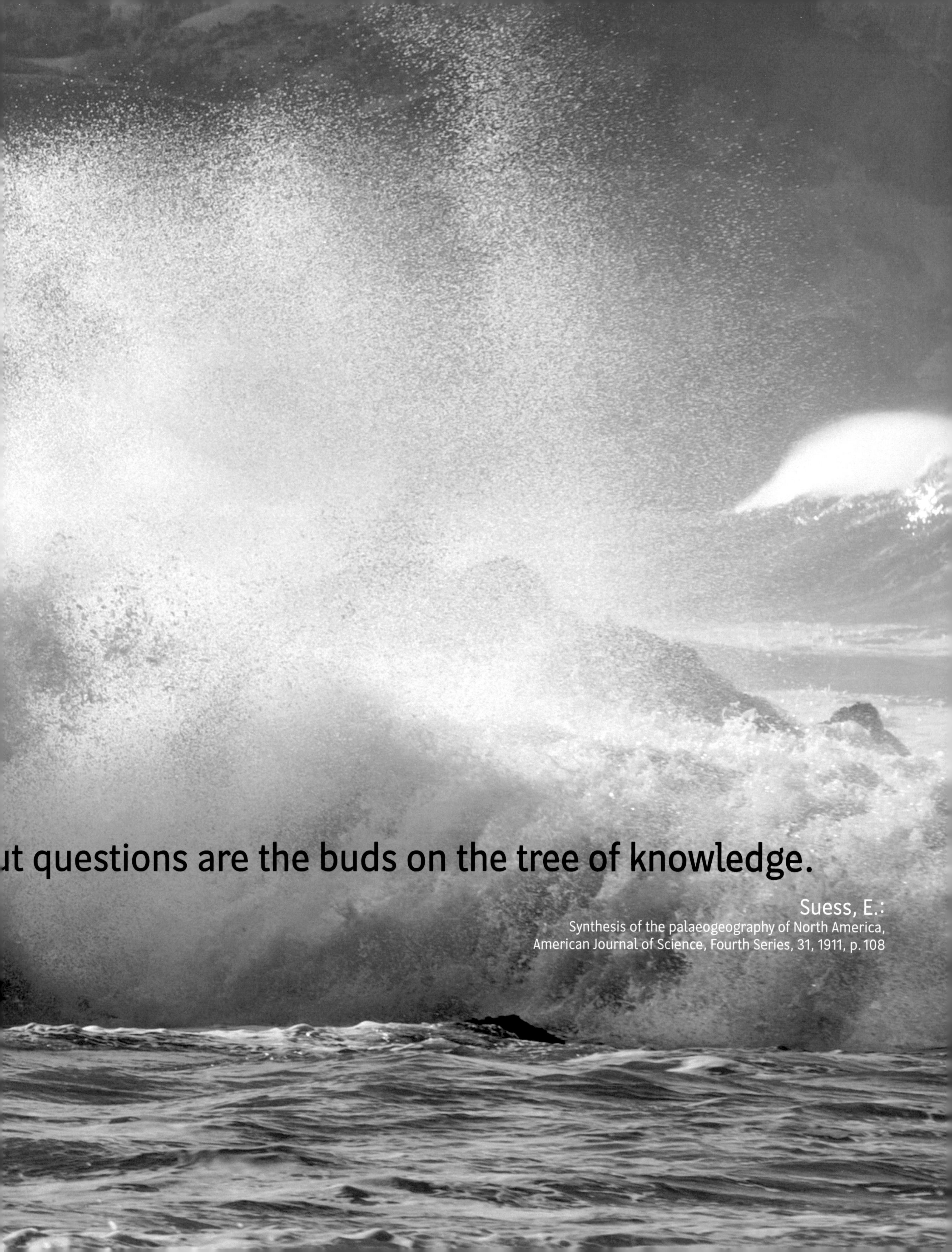

ut questions are the buds on the tree of knowledge.

Suess, E.:
Synthesis of the palaeogeography of North America,
American Journal of Science, Fourth Series, 31, 1911, p. 108

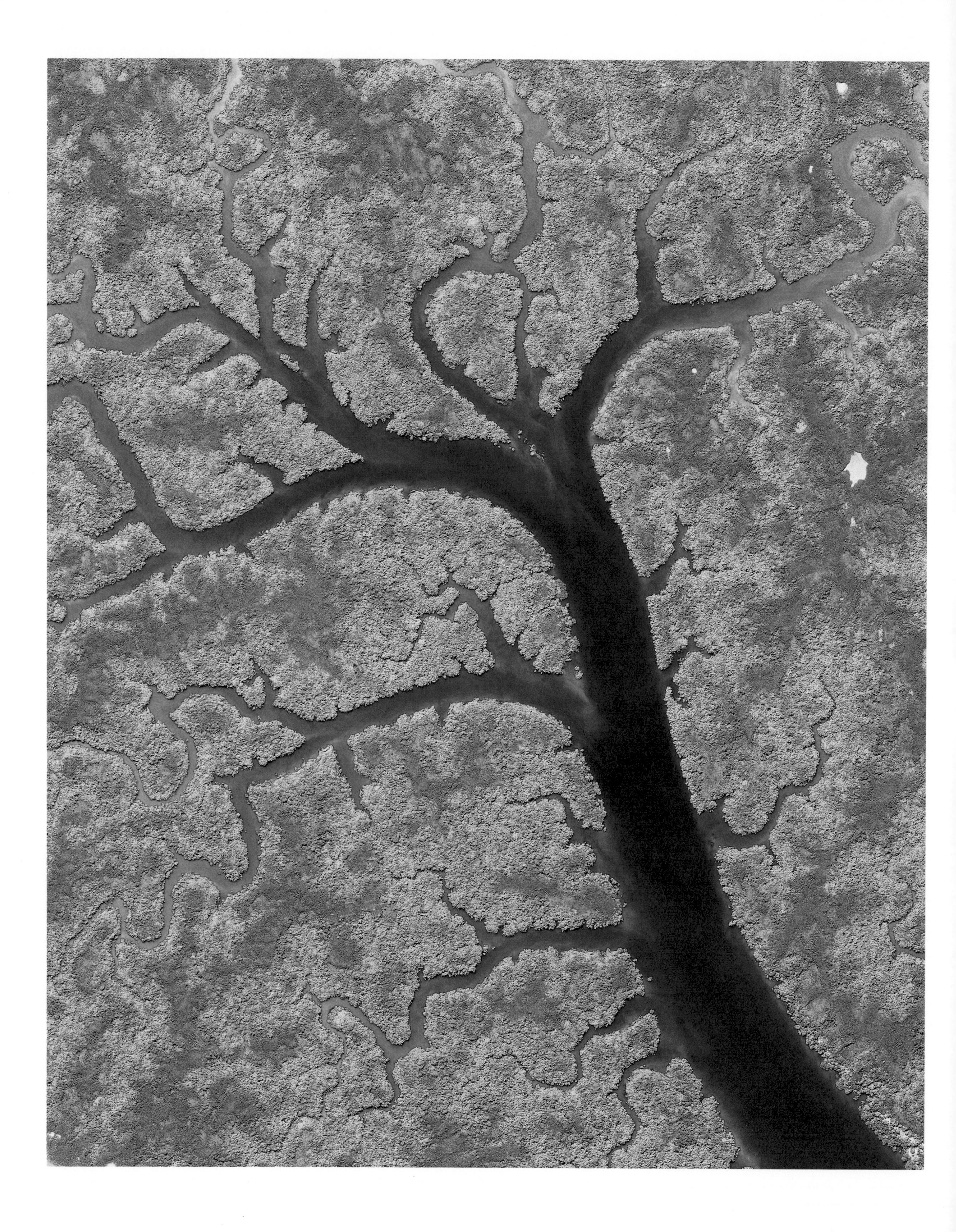

PREFACE

Like a human face, our planet exhibits a large diversity of intricate shapes and patterns. In the dynamic earth system, processes continuously create, modify and destroy specific forms, while at the same time individual forms and larger patterns constrain how processes operate.

It was geologist Eduard Suess who introduced the metaphor of the 'Face of the Earth' to represent the intriguing complexity of the interplay between geoscience processes and forms or, in short, what the earth looks like and why. Indeed, the metaphor of a face has much in common with what we learn about the earth – interesting patterns that have come about for a reason, and the diversity in these patterns may be allegoric for the diversity of geoscientists themselves. It is no coincidence that Suess chose 'The Face of the Earth' as the title for his *opus magnum* which, in essence, is about process and form.

The diversity of geoscience processes and forms is apparent in all spheres. Atmosphere, hydrosphere, lithosphere and biosphere have become common terminology, the last having been coined by Eduard Suess. He also anticipated the importance of the links between these spheres in the earth system and the role humans play in them which have only recently been fully appreciated by the wider community. We are no longer only residents but have become co-designers of this planet.

This book is published at the time of the centenary of Eduard Suess' death on 26 April 1914. It attempts to provide a snapshot of his legacy, what remains today of his vision that is relevant to the theme of the face of the earth in the 21st century.

The book presents photographs from around the world, illustrating processes that have shaped the earth as it is familiar to us today. These facets of geoscience processes are complemented by quotations from the extensive work of Eduard Suess in which he first identified mechanisms or suggested names for concepts that are still valid today. Text, photographs and explanatory quotations are intended as a mosaic of patterns to illustrate the multiple facets of our ever changing planet.

The main body of the book is preceded by five introductory essays. A. M. Celal Sengör, in his first contribution, addresses the relevance of Suess in shaping geology as we understand it today – a science at the crossroads of multiple disciplines. His second article highlights the overarching importance of Suess in the field of tectonics as a formative force for the earth's shape and its ongoing modification. Werner E. Piller, as a palaeontologist and stratigrapher, illustrates how evolutionary processes and environmental changes have always shaped the biosphere, which, in return, has influenced and modified the other spheres from the very beginning. Günter Blöschl narrates two cases of how geosciences can interact with society – flooding and water supply problems of a big city. Finally, Thomas Hofmann draws together the legacy of Eduard Suess who was not only influential in the geosciences but also in society; a genuine European, born in England and living in Prague and Vienna. It becomes clear that what matters is not only to have a vision but also the ability to put that vision into practice.

The authors hope that this volume will contribute to a better appreciation of how our understanding of processes and forms across all spheres of the geosciences is constantly evolving, and how these very processes are shaping the face of the earth.

Thomas Hofmann, Günter Blöschl, Lois Lammerhuber, Werner E. Piller, A. M. Celâl Şengör

Cover | Flysch, Zumaia, Spain
4 – 5 | Flysch, Zumaia, Spain
6 – 7 | Bay of Biscay, Spain
8 – 9 | Supercell thunderstorm cloud, Montana, USA
10 | Saint George Basin, Prince Regent River, Australia

Science leadership. The bust of Eduard Suess in the Aula of the Austrian Academy of Sciences where he was president from 1898 – 1911.

W. FRASS

EDUARD SUESS AND THE ORIGIN OF MODERN GEOLOGY

A. M. Celâl Şengör
Istanbul Technical University

The human mind wishes to understand. It demands complete and reliable knowledge of things and wants to comprehend their workings, but this desire is almost impossible to fulfil. The world in which we live is far too large to be embraced in its entirety by our very limited faculties, even if it were possible to direct to such a task the attention and industry of the entire species. Yet we find it impossible to quench our thirst for comprehending the world around us. This is in part a reflex conditioned by our biological evolution: we know that ignorance and misunderstanding may cost us our comfort, security, even life altogether. Our minds support this reflex: we realize that the more we understand, the better the life we can design for ourselves.

The anxiety resulting from the fear of not knowing and hence not understanding has pushed mankind in two separate directions, long held to be mutually exclusive: 1) a holistic approach, by which our mind tries to grasp the whole by trying to recreate it in itself; 2) a particulate approach, whereby our minds try to grasp the world by considering its parts *seriatim*. Religions and Descartes' rationalism are examples of the first approach, whereas Bacon's attempt to know and understand things simply by listing their properties is the best known example of the second. We know that both failed and yet we still attempt to justify them, because we despair of finding a third way outside our reason and our senses. This desperation stems from our desire for certain, indubitable knowledge.

Mankind's desire to get to know its home, the planet earth, has pushed some of its members along the first and others along the second path. Plato thought, in *Timaeus*, that he could divine the features and the character of the universe by thinking about it, whereas his pupil Aristotle recommended, in his numerous treatises on diverse aspects of nature, that we first observe the universe. He thought induction would then give us understanding, but he realized that simple enumerative induction would not do.

When geology was first coming into being as a discipline of its own, it found a ready-made hypothesis of earth history: the Biblical deluge. This hypothesis had the great advantage of making global correlation of all sedimentary deposits possible by positing two major events in earth history: the Creation (during which all 'basement' rocks had formed) and the Flood, which had allegedly laid down all the layered sedimentary rocks. All subsequent developments in geology until the middle of the 19th century represented attempts to refine the flood hypothesis (by which time it had long been severed from its religious roots) by trying to find ways to correlate the various sedimentary packages and to discover the processes that created the receptacles in which the sedimentary rocks had come to rest and the source regions from which they had been washed down.

When Eduard Suess (1831–1914) began being interested in geology (in 1849, when he was only 18), there were two rival claims for earth behaviour and history. One argued that the earth had a stop-and-go behaviour, generating an interrupted but readily identified record of unerring regularity. This view was sympathetic to those who sought a conformity of the earth history to the one related in the Bible. This was a holistic view and assumed previous knowledge of the manner in which our planet behaved. The other claim held that we actually did not know how the earth behaved and that we could only come to know it by looking at the processes now active and assuming an unchanging, yet irregularly proceeding evolution dominated by the earth's currently observable operations. This view required a detailed knowledge of individual processes acting now and their presumed traces in the earth's past.

Suess' interest in geology had been aroused when still a teenager by his fascination with traces of past life as reflected in the fossils he first saw in the museum in Prague, today's Národní Muzeum on Wenceslas Square, and he found enquiry into the nature of the vanished worlds in which the fossil animals and plants once had lived irresistible. His first job in the Imperial Natural History Cabinet, the ancestor of the present-day Natural History Museum Vienna, led him into palaeobiological studies, starting with graptolites and brachiopods. In these studies one can detect an attempt to understand these fossils as representatives of once living zoological entities and the environment in which they flourished. This inevitably brought him into contact with the rocks that entombed the fossils. Suess' study of the environments required him to look at the world today where similar animals lived and he came to admire the endless variety and uninterrupted continuity of what we would today call ecosystems.

When he looked at the rocks however, he was astonished to find well-delimited beds piled upon beds containing the remnants of past ecosystems that seemed confined to certain individual beds or groups of beds that looked independent of those below and those above them, as Baron Georges Cuvier (1769–1832) and his followers had demonstrated. Earth history appeared to consist of a book of short stories that were conceived independently of one another and the division between the pages of each history was interpreted as a global catastrophe. How was such a thing possible on a planet on which the various ecosystems today appeared to be governed by the same sorts of processes whose products formed the contents of the beds that presented an essentially interrupted record? Sir Charles Lyell (1797–1875) and Charles Darwin (1809–1882) tried to solve this problem by assuming that a large number of pages of the book had later been removed, the interruptions between the stories were more apparent than real and the book of short stories was in reality a long and complete novel. Suess discovered that the claims of Lyell and Darwin were based on local observations and that they had failed to consider global data. When he looked at world-wide reports of the gaps in the record, he found that their horizontal extent was such that they truly required global events to form them and that local removal of parts of the pages of the book of the earth would not do. This observation, however, created a terrible dilemma: how was it possible in a world, where local Lyellian *steady-state* processes seemed to dominate, to generate a record of earth history that looked Cuvierian, i.e. *locally episodic*?

The answer presented itself from a most unexpected corner. Suess had for some time been studying the basin in which his home lay: the Vienna Basin. This structural interruption right in the middle of a majestic train of mountains had formed during the later Cenozoic as a marine basin that later turned brackish and later still entirely freshwater, ending up housing a fluvial system of which the present Danube was the current member. The transitions from one environment to the other appeared abrupt and until 1862 Suess assumed, as a loyal Lyellian, that successive phases of uplift had cut off the basin from the world ocean. To his surprise he discovered, during the course of a correspondence with the Russian geologist Nikolai Pavlovich Barbot de Marny (1829–1877), that the same sedimentary layers as in the Vienna Basin, at the same elevations, could be observed around the Black Sea, north of the Caucasus and all the way to the Aral Sea. Suess thought that no uplift could create such a perfect agreement, suggesting lack of deformation, at such immense distances. The only way to explain the record seemed to him to assume a change in the level of the sea without touching the continent itself. In other words, Suess needed a process to allow him to change the volume of the ocean basins without at the same time moving the land.

The only *vera causa* available at the time was, he thought, the French geologist Constant Prévost's (1787–1856) version of the theory of thermal contraction of the planet. Prévost thought that planetary contraction was spatially irregular and led to local subsidences of large areas, which formed the ocean basins. He thought such subsidences had taken place at various times in the history of the earth and considered mountain-building simply a '*contrecoup*', i.e. a reaction to the subsidence. Such a theory was precisely what Suess needed as a *realis causa*: he thought that any subsidence, *however local and small*, must create an increase in the capacity of the ocean basins and result in a *global* regression, i.e. a withdrawal of the ocean from the continental edges. When the space thus created was later filled again with sediment, the lost volume would be recovered and the sea would transgress its borders onto continents. Because the world ocean is a connected body of water, any local subsidence would create a global regression, and its subsequent transgression onto the margins of its basin must of necessity be global. This simple way of looking at world history had in one blow removed the Lyellian-Cuvierian contradiction: they had been both right. The global record was indeed discontinuous, as Cuvier had rightly claimed, and yet the planet was governed by irregular, local events that happened now here and now there, as Lyell maintained on the basis of what we observe today. Cuvier was wrong in insisting on global catastrophes and Lyell was wrong in denying the possibility of any global event.

The Vienna Basin dilemma thus resolved, Suess turned around to look at the magnificent Alps and the Carpathians surrounding them. He had long realized that they were actually a single system of mountains, later divided by the

Vienna Basin. How did they fit into his new world picture? What, in effect, was the nature of what Prévost called *countrecoups* to oceanic subsidences?

When Suess was an assistant in the Imperial Mineral Cabinet, he also worked as a volunteer in the newly-founded Imperial Geological Survey, the *Geologische Reichsanstalt* in Vienna. In that capacity he had been detailed to assist the veteran Austrian geologist Franz von Hauer (1822–1899) in a mapping exercise between Passau and Duino. Suess, an ardent mountaineer, had requested the highest part of the traverse to be mapped, the gorgeous Dachstein Massif, as his assignment. What he saw in the Alps then and on his previous encounters with geological structures in the Carpathians and in the Bohemian Massif had shown to him that the then prevailing ideas of mountain building could not be right. One assumed that every segment of a mountain range with a uniform strike had formed by a violent uplift along a straight central axis caused by rapid igneous intrusion, helped along by planetary contraction leading to the folding of its crust. One further assumed that mountain building was a planetary event and that all mountain ranges with parallel trends were simultaneous in their origin. The great German geologist Leopold von Buch (1774–1853) had put forward the violent vertical uplift theory to which the French geologist Léonce Élie de Beaumont (1798–1874) had later added the contraction component and the global synchroneity, coupled with an episodic history of mountain building. Élie de Beaumont had in fact said that he had done nothing more than combine Cuvier's global catastrophes wiping out the animal and plant population episodically with Leopold von Buch's violent mountain uplifts. His only contribution had been to introduce the idea that the earth's mountains had formed along the edges of a regular dodecahedron, the shape to which the cooling sphere was tending.

As a stratigrapher, Suess knew that mountain building was not a violent, one-phase process. During his mapping he had proven that the great crystalline massifs of the Alps were as much passive victims of the tectonic events as were the sedimentary parts. His knowledge of the Alps had convinced him that the mountain ranges arose slowly and continuously across many geological stages and they evolved from their interior to their exterior. They were also asymmetric, not only in the Alps and the Carpathians, but also in the Apennines and in the Appalachians, as the Rogers brothers had shown in their superb mapping. Suess became impressed with the German geologist, botanist and poet Karl Friedrich Schimper's (1803–1867) idea that the vertical uplift theory of the Alps was untenable and everything seen in their structure indicated horizontal shortening. In his somewhat polemical paper, presented in Erlangen in 1840, Schimper had proposed that a thermal contraction of the planet could account for the shortening deduced from the Alpine structures. This was along the lines of Prévost's ideas and Suess thought it the best theory available. James Dwight Dana (1813–1895) in the United States was trying to combine the ideas of Élie de Beaumont with the findings of the Rogers brothers into a comprehensive theory of earth contraction, in which the oceans appeared as broad synclines and continents as broad anticlines. Following Élie de Beaumont, Dana believed mountain ranges were anticlinoria that had grown out of previous synclinal troughs that he called geosynclines.

Suess knew that this view could not be correct, simply because he believed the basic premises of Élie de Beaumont's hypothesis to be wrong. Instead of adopting a polemical attitude to defend his own views, Suess chose to collect global data to test his own slowly emerging synthesis: he thought orogenic unconformities local, but the major transgressions and regressions global. Geologists had managed to build a global geological timescale that was valid everywhere simply because the trans- and regressions created global sequences. These sequences contained their peculiar fossils, making world-wide biostratigraphic correlation possible, because regressions reduced the shelves, destroying much of their rich marine life, and altered the climate and physical geography of the lands, leading to abrupt changes in the terrestrial biota. Transgressions introduced new life forms by accelerating the biological evolution that resulted from enlarged marine living spaces.

Suess had formed his view of earth behaviour and its history by the early 1870s. One would have expected him to present it to the scientific world like most of his predecessors had done: as a compact and complete theory in articles in scientific journals and eventually in a textbook. He did not do that. Instead he decided to examine the face of the entire planet, sifting a vast amount of geological data from the literature with a view to testing his theory. This procedure eventually led him to write two great books, the slim volume *Die Entstehung der Alpen* (The Origin of the Alps) in 1875 and the massive, four-volume *Das Antlitz der Erde* (The Face of the Earth), between 1883 and 1909. This worked in his favour as an author, his staggering erudition earning him an immense reputation as the dean of world geologists and unbounded admiration. Yet it worked against his theory, because very few people seem to have read either of the books carefully or thoroughly. They were considered regional descriptions, a reflection of the face of the planet, and one eminent contemporary admirer went so far as to say that Suess claimed nothing in his books, but simply showed. Because he was an advocate of the contraction theory, he was grouped with other prominent contractionists, such as Élie de Beaumont and Dana, and his *Antlitz der Erde* came to be viewed as a rich repository from which world regional data could be quarried. Many laughed at his insistence that mountain-building was created by a one-sided push, because it was thought that he had not understood simple Newtonian mechanics, in which every push created an equal and opposite reaction. Élie de Beaumont's simile of mountain-building with the workings of a vice was thought a more realistic analogy than Suess' talk of waves breaking on a beach. It was pointed out that waves generate no horizontal motion and that Suess' metaphor was ridiculous. It was overlooked that Suess' asymmetric mountains had various levels of detachments underlying the entire mountain range and that the asymmetric push had a very real existence, because the 'reaction' was spread out right across the mountain range along detachments that explained the decrease of deformation from the interior parts to the exterior. It was ignored that waves, when they break on a beach, carry material forward.

But the problem with Suess' writings was simply that they demanded an immense effort on the part of the reader. They are not sermons, but invitations to debate. His book is not a reflection but a *detailed reading* of the face of the earth. His interpretations are interspersed across long and detailed regional descriptions, every word of which is necessary to understand the interpretation. In reality Suess had adopted Charles Darwin's style of the 'long argument' to present his idea of the planet crumbling under contraction. Like Darwin he long withheld its publication and when it appeared in print, it looked like a vast repository of observations. Its presentation required the reader to have a geographical atlas at hand and paper and pencil ready to take notes and scribble sketches. Suess' books are not books to be read by the bedside.

Only one geologist after Suess really read his work as it demanded to be read and understood it fully: Émile Argand (1879–1940). The resulting *La Tectonique de l'Asie* (Tectonics of Asia, 1924) is a richly and beautifully illustrated culmination of Suess' thinking minus his rejection of the hypothesis of isostasy, i.e, a presumed gravitational equilibrium between the earth's lithosphere and its mushy underbelly. In many ways, Alfred Wegener's (1880–1930) theory of continental drift is also along the lines of Suess' tectonics; Argand showed how it accommodated itself in a Suessian world. It was a world where all available observations were subjected to continuing and comprehensive tests. This was a holistic approach checked by *seriatim* tests. It presented no final truth, but a way to approach the truth asymptotically. It presented no ready-made dogma, but an invitation to an endless journey of exploration. After Argand withdrew from geology, nobody conducted geology as Suess had recommended it until J. Tuzo Wilson (1908–1993) came forward with his theory of plate tectonics in 1965. The period that intervened between 1924 and 1965 was thus a dark intermezzo, when geology became mundane, local, uninspiring and began disintegrating into individual disciplines that tried to emulate physics, chemistry and biology. Wilson's return to Suess' way of doing geology shook the geological community and reminded it of the oneness of the planet. Modern geology as we know it today is thus largely a creation of Suess and his few followers.

What is a geological formation? What conditions determine its beginning and its end? How is it to be explained that the very earliest of them all, the Silurian formation, recurs in parts of the earth so widely removed from one another – from Lake Ladoga to the Argentine Andes, and from Arctic America to Australia – always attended by such characteristic features, and how does it happen that particular horizons of various ages may be compared to or distinguished from other horizons over such large areas, that in fact these stratigraphical subdivisions extend over the whole globe?

The Face of the Earth Vol. I., Clarendon Press, Oxford, 1904, p. 8

Transgressions. The transgression of Miocene strata on crystalline rocks of the Bohemian Massif in the vicinity of the city of Eggenburg, Lower Austria. Original illustration by Suess, about 1860.

FROM PALAEONTOLOGY AND STRATIGRAPHY TO EARTH SYSTEM SCIENCE

Werner E. Piller
University of Graz

Suess started his academic career as a palaeontologist. His first scientific publications dealt with fossils focusing on graptolites, brachiopods and ammonites but also on vertebrates. He placed special emphasis on brachiopods covering a time span from the Palaeozoic to the Cenozoic. Although describing a great number of new species, Suess was primarily not a taxonomist but a palaeobiologist. This means that he not only described and categorized fossils but looked at their extant equivalents. He was even in favour of integrating soft part morphology of living representatives of an organism group to understand the functional morphology and ecological requirements of extinct creatures. This clearly indicates that Suess was, from the very beginning of his career, interested in understanding processes and interrelationships of overriding importance.

His palaeontological work has to be seen against the rivalling cataclysmic views of Cuvier and Lyell's strict uniformitarian approach. Suess was fully in line with the actualistic view of Lyell to explain most of the geobiological processes on earth. He even accepted the idea that more ancient biota show greater differences from modern ones than more recent communities. He actually regarded Lyell's quantitative approach positively – comparing faunas of different ages on the basis of the similarity expressed in percentages. However, he also clearly pointed out that this simple principle cannot be applied invariably. One of his most evident examples in this context was the Miocene Sarmatian fauna (see below). This fauna deviates almost completely from that of the preceding stage (nowadays called the Badenian) in sharing only a very small number of taxa. Suess interpreted this striking difference as deriving from a major tectonic, palaeoceanographic and biogeographic reorganization in the area of the Paratethys (a term introduced, however, much later by Laskarev in 1924).

The distribution of biota on different spatial and temporal scales was one of the major basic principles which Suess clearly addressed in his early work. When dealing with the habitats of brachiopods, he mentioned barriers, such as mountains and seaways, restricting horizontal distribution, and climatic impediments being most important for the temporal distribution. As an additional important factor for large-scale palaeobiogeographic distribution patterns, Suess cited transgressions (sea level rise) and regressions (sea level fall) acting as isolating or connective agents which consume or create living space. On a basin scale, frequently referring to the distribution within the Vienna Basin, he clearly demonstrated that distinct distributional patterns exist, evoked by facial differentiations, e.g. coastal versus deeper water environments. As a consequence he concluded that fossils are much better environmental indicators than sediment or rock characters.

Sea level changes and their origin was one of the key topics during his research. Falls of sea level, called regression, could be observed at numerous locations across the world through the entire earth history. These changes had been allocated by earlier authors to so-called secular elevations and depressions of entire continents. In contrast, Suess explained them by a regional subsidence or structural collapse of the sea floor, also connected with the formation of large ocean basins. These processes create a global lowering of the sea level and consequently global regressions. Suess called these sea level changes *eustatic movements* and concluded that regressions are produced by *spasmodic eustatic negative movements*. This phenomenon of a changing eustatic sea level is in fact the basic mechanism for what we nowadays call sequence stratigraphy. This allows sea level changes to be correlated on a global scale over various magnitudes as well as on a basin scale, representing a widely used stratigraphic correlation tool. For Suess, subsidence was the only, or at least, the key agent. He did not take into account the waxing and waning polar ice caps, currently considered equally or even more important for sea level changes.

Eustatic sea level changes enabled Suess to explain the laterally extremely wide distribution of rock successions, as well as major gaps, on a continental and even global scale. These sea level changes create not only major transgressions, which increase inundation of land areas, but also major regressions that cause a tremendous amount of weathering and erosion. By studying these phenomena on a global scale, he was able to infer major sedimentary cycles and to reconstruct the palaeogeography of particular areas of the globe for different eras. Based on these results Suess was able to reconstruct the palaeogeography of North America long before Lawrence Sloss presented his *Tectonic Cycles of the North American Craton* in 1963. In this later paper these cycles were called 'Megasequences' and were later considered the inception of modern sequence stratigraphy – totally ignoring the oeuvre of

The world's largest fossil oyster reef, which formed 16.5 million years ago
in a tropical estuary and was destroyed by a tsunami. Stetten, Lower Austria.

Suess who had already come to a very similar derivation and realized to which extent these processes shaped the 'face of the earth'.

Sea level changes naturally also affect biota and Suess asserted that even small-scale changes are of great importance. They influence shallow marine organisms in particular, while land biota and deep sea fauna remain unaffected. An outstanding example of the influence of sea level changes on a widespread fauna are the very fossiliferous *Cerithium* beds of the Vienna Basin which can be traced eastwards as far as the area of Lake Aral. These beds are well characterized by a peculiar, low-diversity molluscan fauna (gastropods and bivalves), rich in individuals, which is unique in its composition over this huge area. This fauna was interpreted as reflecting not only very specific environmental conditions but also a particular time span. In contrast to earlier opinions, Suess pointed out that it was not the gastropod genus *Cerithium* that was the critical key taxon but several bivalve taxa, such as *Mactra* and *Ervilia*. The gastropod *Cerithium* is a very abundant faunal element but not diagnostic for these beds. Earlier, Suess had interpreted this fauna as the result of reduced salinity. He called these deposits the 'brackish stage of Vienna'. Later on, however, Suess realized that a great variety of environments of variable salinities is represented within the sediments of this stage. Due to this new interpretation and spatial distribution he named these beds the Sarmatian stage and the fossils represented the Sarmatian fauna. This stage is still in use for the regional chronostratigraphic classification of the Paratethys as the upper part of the Middle Miocene. In many recent papers, however, the old idea of the brackish character of this fauna is still upheld, ignoring the revised view of Suess.

Another very important and crucial incident in the early phase of Suess' scientific activity was the publication of the *Origin of species* by Charles Darwin as well as the preceding ideas of Lamarck. The genetic relationship between all living organisms was a given fact for Suess and the search for evolutionary lineages and the establishment of a natural biological system the declared goal. The aspects of gradualistic evolution and selection held a central position in Suess' (palaeo)biological and stratigraphic reflections. In the course of his own studies and his compilation work, it became very obvious for Suess that the fossil record supports Darwin's (and also Lamarck's) ideas to a certain extent. Suess, however, already realized that some taxa show a co-evolutionary trend leading to – what we call today – the evolution of ecosystems (called 'complete economic units of Nature' by Suess). In addition, Suess stated that evolutionary changes do not necessarily develop coevally but differ widely, depending on changing environmental parameters. This divergence is clearly expressed when comparing, for example, terrestrial with marine faunas. Although these differences do exist, it appears clear on a global scale that biota have a certain characteristic for particular geological periods and can therefore be used for biostratigraphic correlation. In combining palaeobiology and stratigraphy, he clearly realized that organic remains are the primary and most important tools for understanding earth history. In addition and somewhat in opposition to Darwin's opinion, he insisted on the very important influence of environmental changes on evolution. While he did, to some extent, agree with Darwin's gradual evolutionary paradigm in stating that the development of life was non-interrupted, in his view it was not continuous. For Suess the fossil record needed an additional mechanism to explain the relatively abrupt and global changes in biotic composition. Suess advanced the view that the origination of a species is a rapid process compared with its long existence without any change. With this view he was, in fact, in opposition to Darwin and presented an already highly punctualistic evolutionary model. As a driving force for evolution he cites several parameters (e.g. climate, oceanic water circulation, humidity), however, he considered the changing sea level the most important influencing factor, not only for the spatial distribution of biota but also for their evolution. This integration of palaeobiology, evolution and stratigraphy strongly demonstrates his holistic view to understand the 'face of the earth'.

As a result of Suess' global geological view, he invented the terms Gondwanaland and Angaraland. Gondwanaland was biogeographically defined by the occurrence of a specific flora – the Gondwanda or *Glossopteris* flora – and includes a major part of Africa, Australia, Madagascar, the Indian Peninsula and South America. Angaraland included the northern landmass also characterized by a distinctive Permo-Mesozoic flora. Although Suess did not know about the mechanism (Wegener's mobilistic ideas were formulated later), he inferred these large land masses merely on the basis of their biotic distribution. This knowledge was based on very thorough analyses of lithology and stratigraphy covering both areas to a wide extent.

Furthermore, the ancient continent of Gondwana was framed by a big ocean, which Neumayr had named 'Centrales Mittelmeer' (Central Mediterranean) earlier, but which Suess called Tethys. He did not just introduce this name, but characterized this ocean by its geological features over its distributional area and stratigraphic range. Finally, he considered the actual Mediterranean Sea as the remnant of this ocean. Years before Wegener's idea of continental drift and the resulting hypothesis of plate tectonics, Suess interpreted oceans as geologically not necessarily long lasting units but as changing entities. He realized that Asia originated from the disappearance of the Tethys Ocean and the fusion of Angaraland with the Indian fragment of Gondwanaland. Although embedded in a static view of the world, he asked, "... why we should believe that all the great ocean basins have been continuously covered by water since panthalassic times", expressing the view that oceans are changing and even vanishing entities. Comparing the Atlantic and the Pacific, Suess also realized that the margins of these oceans are fundamentally different: the Pacific is framed by mountain chains which are missing along the Atlantic rim. Again, although he had no knowledge of plate tectonics, he clearly identified active and passive continental margins. Suess considered processes related to ocean basin formation to be responsible for the changing shape of the earth surface. In close dependence on these processes, Suess repeatedly pointed out that the biogeographic distribution of biota, mostly demonstrated by terrestrial plants and animals, can undoubtedly be explained only by these processes. Contrary to Alfred R. Wallace, who interpreted the origin of many islands by uplift, Suess only accepted subsidence as the possible mechanism to explain the great similarities between islands and distant continents, but was aware that drift of biota can be an additional mode of distribution.

Suess approached his science as a palaeontologist who did not remain with taxonomy but attempted to consider (fossil and recent) biota from an ecosystem perspective. Through combining the great variety of his observations and inferences, Suess became aware of the interrelationships between biotic activities. He considered all biota which represent 'the totality of the animated earth and which live above the lithosphere' as belonging to the biosphere. The biosphere was defined by him as a 'spatially and temporally restricted living entity'. Within the biosphere he separated two major groups of biota, one living in the sunlit ('insolated') part of the world, the other in the dark sphere. He also repeatedly focused on the fact that animals settle in very different habitats (land animals, seawater and freshwater animals) and that they are in a distinct food dependence (herbivores vs. carnivores), thereby explicitly addressing the topics of the food web and trophic chain.

Beyond the biosphere he described several overarching entities which are wrapped around the earth like envelopes and which interact in a complex way with each other. The atmosphere was already a traditional, well established entity, representing the outermost envelope, and he certainly saw it as part of the whole system. He assigned all rocks of the planet to the lithosphere and he started to look at both the litho- and biosphere to understand their interconnection and mutual interference. In addition to the litho- and biosphere, he named the ocean water the hydrosphere. Its interaction with the other spheres is obvious and manifold. For example, while changes in the hydrosphere account for the origination and termination of epicontinental seas, those of the lithosphere are responsible for changes of deep ocean basins. This statement again refers to the importance of transgressions and regressions for shaping of earth's surface, for changing living space and affecting the biosphere and the transience of ocean basins. Although the meaning of the three spheres introduced by Suess is not fully in line with their current use, Suess pointed out very clearly that the distribution of biota in the oceans, for instance, depends on temperature and light besides their other requirements and that all spheres are tightly interrelated. Contrary to many other coeval natural scientists, Suess' intellectual world roots in a geological tradition that allowed him to understand that spatial environmental distribution changes through time and consequently also the interrelationships between the spheres. This holistic approach, combined with his interdisciplinary thinking, brought him close to what we would today call Earth System Science.

Finally, the open minded and anticipatory view of the ingenious personality of Eduard Suess is best rendered by the quotation: "Our scholars will someday know more than their masters do now; so let us patiently continue our work and remain friends."

Das beste Bild, das ich von der Entstehung grosser Gebirge zu geben weiss, besteht darin, dass ich mir vorstelle, es würde meine Hand durch irgend eine Verletzung aufgeschürft, dabei die Haut nach einer Seite in Falten zusammengeschoben, auf der anderen reisse sie und es dringe etwas Blut hervor. So sehen wir grosse Gebirge immer nach einer Seite zusammengeschoben, in grosse Falten gelegt, – und diese Falten ziehen sich hier durch das ganze Königreich Sachsen, – auf der anderen Seite zerreissen sie, und wo sie aufgerissen sind, da treten ans dem Innern der Erde Vulkane hervor.

The best image I can offer about the formation of large mountain ranges is to picture an injury to my hand in which the skin is pushed in folds to one side and ripped open on the other, with some blood oozing out. In the same way we see large mountain ranges pushed together to one side in large folds – here these folds run through the entire kingdom of Saxony – on the other side they rip, and where they rip, volcanoes rise from inside the earth.

Die Heilquellen Böhmens, Alfred Hölder, Wien, 1878, p. 5

Tectonics. The Swiss Alpstein Massif with its highest peaks, the Säntis (left, 2 501 m) and the Altmann (right, 2 435 m) and Lake Constance (left background). The steeply inclined folds of Mesozoic limestone are part of the Säntis-Drusberg Nappe.

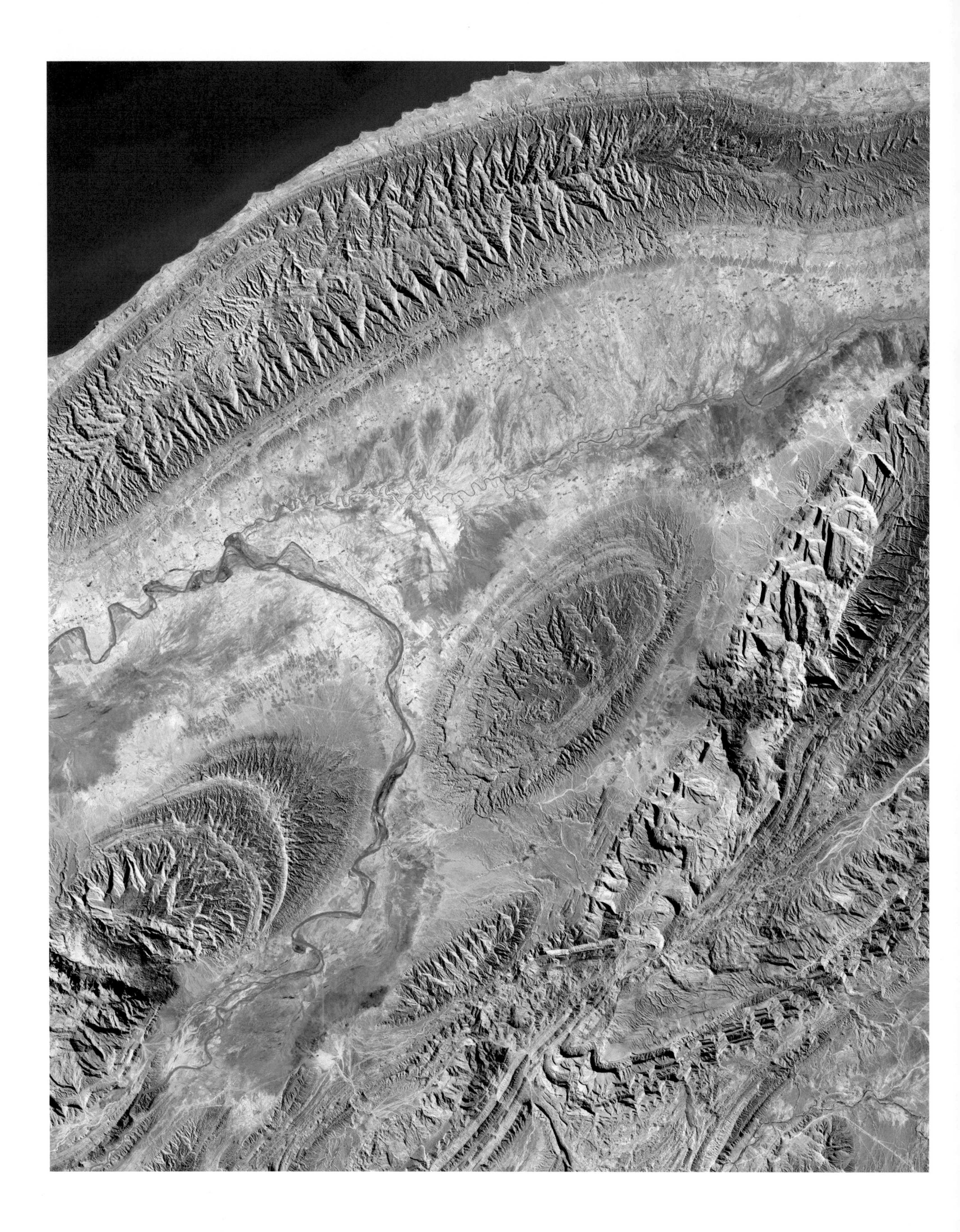

SUESS AND THE DYNAMICS OF THE PLANET EARTH

A. M. Celâl Şengör
Istanbul Technical University

Eduard Suess is possibly the most widely-quoted author in the history of tectonics, but most people who quote him are not aware of the fact that they are quoting his work: they use his terms and concepts, but they do not know in what contexts these were introduced and by whom. The word tectonics was borrowed by the German geologist Carl Friedrich Naumann in 1850 from the ancient Greek word *tektonikos*, meaning carpenter, builder or master in construction, derived from the root *tekton*, i.e. a master in any art. By *Geotektonik*, Naumann designated the study of any kind of structure in rocks, including those resulting from sedimentation and igneous action. Tectonics is now considered to be that branch of geology which studies the deformations of the earth's rocky rind or lithosphere, as expressed in its structures and the motions creating those structures, which arise from flow within the mantle underlying the earth's lithosphere. All of this is a consequence of the fact that the earth is a cooling body. With few exceptions, it was believed until the 20th century that the earth lost its heat by conduction and radiation, and tectonic theories to account for the deformations of the lithosphere were based on that assumption. Only in the 20th century was it realized that convection in the mantle is the main force driving the surface motions we see expressed in such structures as continents and oceans, mountain ranges and rift valleys and large strike-slip faults. With the exception of those who accepted Wegener's theory of continental drift, most geologists prior to plate tectonics tried to accommodate a modest amount of convection in the mantle within the context of a planet losing its heat, mainly by conduction. Only after the rise of plate tectonics was the dominance of the convection currents in the earth widely accepted.

When Suess entered geology shortly after the middle of the 19th century, geologists were still recovering from the shock that Hutton (the man who made it a principle to deduce the past of the earth by familiarizing oneself with its present-day processes) and Sir Charles Lyell (the great advocate of uniformitarianism) had delivered to the complacent community by showing that Werner's view of the earth was untenable. The earth of Abraham Gottlob Werner was dead and all the phenomena that geologists could see were thought to result from the retreat of a once universal ocean (an idea that is called neptunism). Hutton showed that the earth had an interior heat (as some had also supposed earlier) and that it manifested itself in making igneous rocks that not only deformed other rocks while ascending to the surface, thereby creating uplifts, but also metamorphosed them into other kinds of rocks that were mostly crystalline. Georges Cuvier had shown that it was possible to arrange the fossil-bearing sedimentary rocks into a relative chronological order and Leopold von Buch asserted that the large mountain ranges owed their existence to catastrophic intrusions along their axes.

It was the French geologist Léonce Élie de Beaumont who weaved the ideas of Cuvier and Leopold von Buch into a coherent theory of global tectonics in the years 1829 and 1830. Using the Kant-Laplace theory of the origin of the solar system, Élie de Beaumont argued that the earth had to be cooling. He thought the outer part would cool faster than the interior and at one point would cease cooling, but the interior would continue to cool and therefore contract. Eventually, the outer, already cooled, layer would be too large for the interior and would collapse onto it, creating folds. Élie de Beaumont thought that the shortening of the outer, already cooled, layer would take place in two stages. First, large folds, of the size of continents and oceans, would form, representing large anticlines and synclines, respectively (he called them positive and negative *bosselements*). The bottoms of the synclines, having descended into the hot interior, would be weakened and would episodically collapse to create closely spaced folds that we now see as mountains. He thought such events would be global and catastrophic, accounting for Cuvier's world-wide cataclysms wiping out most of the world's biota. He also thought that all mountains formed along great circles and that the terrestrial globe was attempting to acquire the shape of a pentagonal dodecahedron as it cooled and the main mountain ranges were placed along the edges of such a dodecahedron.

This was a brilliant, if overly mathematical, theory, but Élie de Beaumont had not entirely freed himself from Leopold von Buch's uplift interpretation. He thought that while mountains formed, large intrusions also happened and defined their axes. He also believed, following von Buch, that large basaltic volcanoes owed their structure to axisymmetric elevation and not to accumulation. Their craters were a result of extensional stresses; they had supposedly nothing to do with eruptions.

Intensely folded, well bedded sedimentary rocks in the Zagros Mountains, southern Iran. The folds (anticlines) are cut open by erosion, exposing their internal structures.

It was with the controversy over the 'craters of elevation' that the downfall of Élie de Beaumont's theory began in the 1840s. His fellow countryman Constance Prévost corroborated – as a result of his study of the Italian volcanoes (including the ephemeral Graham Island that appeared and disappeared in 1831 in the Sicily Channel) and those crowning the Massif Central – what Lyell and George Poulette Scrope, a specialist in volcanism, had said earlier: there were no craters of elevation and all volcanoes owed their volume and shape to accumulation of erupted material. Prévost also objected to the uplift theory of the big mountain ranges, but in Europe most geologists (with such notable exceptions as the Alsatian geologist, botanist and poet Carl Friedrich Schimper) still considered them products of vertical uplift along long fissures. The Alps, with their crystalline axis flanked by unmetamorphosed, but deformed, marginal zones in the south and in the north, were the showpiece of this idea, so impressively illustrated by the great Swiss geologist Bernhard Studer in his classic *Geologie der Schweiz* (1851, 1853). The prolific Freiberg professor Bernhard von Cotta did the same for the mountains of Germany in 1851.

In North America, however, the meticulous mapping of the southern Appalachians by William Barton Rogers and his brother Henry Darwin Rogers in 1843 had shown that these mountains possessed a highly asymmetric structure. It looked as if the entire mountain range had been shoved to the northwest with respect to an undeformed continental interior. They thought a series of violent eruptions in the present-day Atlantic Ocean might have been the causative mechanism, but their countryman James Dwight Dana thought otherwise. He combined Léonce Élie de Beaumont's theory of contraction with the observations of the Rogers brothers and in 1847 claimed that chains of mountains were asymmetric and that they surrounded a continent towards which they had been pushed.

The next decade, the 1850s, was when Suess' geological activity commenced. He proved to be a versatile geologist from the beginning. He started out as a palaeontologist, but liked to investigate the geology of the localities from which his fossils had come. If they were close by, he went into the field to look at the rocks himself. If they were far away, he consulted the literature (his ability to command the international geological literature remains unsurpassed to this day). In one of the earliest mapping exercises he undertook as a volunteer in the Imperial Geological Survey, he was surprised to see how closely the Northern Calcareous Alps resembled the Southern Calcareous Alps (what we today call the Southern Alps) in their stratigraphy. He and his supervisor Franz von Hauer initially thought this perfectly explicable if the central crystalline massifs rose as igneous intrusions and burst a formerly continuous calcareous cover. Suess soon realized, however, that this could not be: wherever crystalline rocks were seen in contact with the calcareous Alps, the contact was not intrusive. It seemed that the crystalline rocks were older and that they had been deformed together with the sedimentary rocks. They could not therefore be the agents of the deformation.

Shortly afterwards, Suess' stratigraphic studies led him to reject Lyell's theory of continental uplift, because the Neogene stratigraphy of the Vienna Basin appeared to be applicable to vast areas from the Alps to the Aral Sea. Suess thought that only a change in the level of the sea could bring about such perfect correspondence over such immense distances. He was aware that the French geologist Constant Prévost had rejected the theory of uplift for volcanoes as well as for mountain ranges. Prévost was a contractionist, but his theory of contraction was very different from that of Élie de Beaumont and Dana. He allowed large areas to founder along steep faults to form the oceans, resulting in regressions, and he believed mountain-building was a simple reaction to this subsidence. Unfortunately, Prévost never specified what sort of a 'reaction' would create the mountain chains of our globe.

It was clear that horizontal shortening of some sort would do the job, but the nature of that shortening and its relation to the observed structures of the mountains had yet to be worked out. Dana had simply adopted the Rogers brothers' superb mapping of the Appalachians, but tried to interpret it from Élie de Beaumont's viewpoint, although Élie de Beaumont had pointed out that his theory would produce symmetric mountain chains.

In 1867 a flooding disaster in the world-famous Wieliczka salt mines south of Cracow occasioned a visit from Suess. Suess realized that the salt body there formed the core of an anticline right in front of the Carpathians. When he looked west in front of the Alps, he remembered that the Mt. Salève anticline near Geneva sat in a comparable position in front of the Alps and a similar anticlinal structure had been mapped by the Swiss and German geologists almost as far west as south of Munich! North of Vienna were the flysch outcrops, which Suess thought to belong to the same anticlinal structure. He suddenly realized that a continuous anticline rimmed the Alps and even the northern Carpathians from one end to the other. To him this suggested that the entire mountain range had been moved northwards. This agreed perfectly with the older observations of the Rogers brothers from the Appalachians, but there was no ocean to the south to do the pushing, as Prévost and Dana had assumed.

It was at about this time that Suess had begun to organize student excursions to southern Italy, where the richness of geological curiosities constituted an excellent teaching venue. In Italy he noticed how similar the Apennines were to the Southern Alps, but they lacked a counterpart to the Northern Alps. For a time he thought Sicily may have been such a counterpart and the Calabrian Massif, plus the Peloritani Mountains, the remnants of what remained of the central crystalline chain. Suess was happy to see that the central crystalline chain had disappeared and he simply assumed that it had foundered. But further consideration of the Apennines, together with the Alps and the Carpathians, finally led him to reject the idea of the symmetric structure of mountain chains. He now thought they were all asymmetric and while their outer rims were shoved outward, their interiors broke up, subsided and generated volcanicity. He looked at other European mountains, all the way north to the British Isles, and noted that they were all pushed north. Although north of the Alps the main movement had ended by the Permian, there were what he called posthumous folds and thrusts that deformed the Mesozoic and even Cainozoic strata, as in the Paris Basin or the Isle of Wight. Suess suddenly realized that the entire European lithosphere was on the move. He resolved to look at all the mountains of the globe to come to grips with this baffling observation and to see whether the 'plan of the trend-lines' of the tectonic features of the face of the earth might yield a key to its internal processes.

On 17 July 1873, he presented a paper to the Imperial Academy of Sciences in Vienna entitled *Über den Aufbau der mitteleuropäischen Hochgebirge* (On the Structure of the Middle-European High Mountains). It is impossible to understand Suess' other writings on tectonics, including his two major books, *Die Entstehung der Alpen* (The Origin of the Alps, 1875) and *Das Antlitz der Erde* (The Face of the Earth, 1883–1909) without a knowledge of what is said in this remarkable paper. At the end of it Suess stated: "The author has reached the conclusion that the entire surface of the earth really moves slowly and inhomogenously, which, between the 40th and the 50th parallels in Europe, is directed north- and northeastward. The so-called old mountain massifs thereby move more slowly than the regions lying between them that pile up and build mountain chains. In Central Europe they generate regular folds on their polar side and tears on their equatorial side."

This implied an autonomous motion of parts of the outer rocky rind of the planet with respect to their neighbours. The way Suess thought he could make such a picture compatible with planetary contraction was to assume different depths of *décollements* under the mountains, whereby the upper layers would be locally attached to the lower, still contracting, parts in an irregular fashion. This would enable him to generate shortening on one side of the upper layer and simultaneous extension at its back. This, together with the asymmetry of the internal structure of the mountain chains, led to the definition of a foreland and a hinterland. Suess noticed that the basins immediately in front of mountains were highly asymmetric, their inner parts belonging to the mountain, but their outer part belonging to the foreland. He called such basins foredeeps and pointed out that they were almost always bereft of volcanism. By contrast, basins behind the mountains had irregular outlines defined by steep, extensional faults. Such basins very commonly had active volcanicity associated with them. Suess was able to recognize the same pattern from the Alps through the Italian chains, the Carpathians, all the way to the island festoons in the Pacific. He noticed that the map view of mountains and the island arcs resembled those of waves. They crowded in front of what seemed to be obstacles and receded from one another where no rampart hindered their progress. Mountain waves branching out from a crowded stem defined what he termed virgations. If two cusps of wave fronts met at a sharp angle, as at the western end of the Himalaya, he called them syntaxes. Sometimes such wave fronts seemed to cross and override each other at what is termed linkages. Suess realized, as a result of his studies of the map view of the deformations at continental scale, that the entire lithosphere looked as if it was flowing and he compared this movement to that of the glaciers.

Suess' view of mountain building was of necessity in contrast with all earlier theories. His stratigraphic work, together with Franz von Hauer in the 1850s, had shown him that the Alps had been built through a long series of tectonic events and that this was still going on, as revealed by earthquakes. A global tectonic hiccup of the kind envisioned by Élie de Beaumont (and later resuscitated by Hans Stille) could not be corroborated anywhere. Wherever Suess looked, he saw continuous mountain building across long geological times. What interrupted mountain building was almost always the occurrence of a nearby subsidence on a grand scale, forming ocean basins or very large depressions. He originally thought that folding, faulting and intrusion during mountain building stiffened the crust, hindering further mountain building, but his study of Asia showed him otherwise. In Central Asia, he was surprised to find mountains that possessed neither forelands nor hinterlands, except where bordered by large undeformed areas at the margins of a subcontinent-sized area of folding. He noticed that the mountain ranges of Central Asia, from what he called the Angaran tableland to the Indian shield, possessed no individuality and, if regarded separately, their origin remained unintelligible. Suess also noticed that these mountains, which he simply called the Altaids, consisted almost entirely of schists, serpentinites, mafic rocks, clastic sedimentary rocks, with subordinate limestones and granites. There were few outcrops of gneisses, in sharp contrast to other lofty ranges of the planet. He compared them to the waves of the open ocean as opposed to chains like the Alps that had a well-defined foreland, which reminded him of waves breaking on a beach. Suess believed that such an analysis of the stable parts and the areas affected by folding on the face of the earth might yield some clues to earth history.

The mountains in Europe appeared to be truncated abruptly at the edge of the Atlantic Ocean, and across the ocean in North America a similar thing was seen with respect to the Appalachians, which stopped abruptly in Newfoundland. In Africa, the Atlas did the same, as did the Cape Mountains further south. In South America, Sierra de la Ventaña seemed equally truncated by the margins of the ocean. No marine sedimentary rocks were seen along the Atlantic shores that were older than the Jurassic. It looked as if a piece had been cut out of a continent that once united the Old World with the New to make the Atlantic Ocean. In the Pacific, by contrast, not only the mountain chains paralleled the coast, but seemed somehow influential in shaping the margins of the ocean. Suess noted that the Pacific marginal chains were thrust towards the ocean and had deep trenches in front of them. Those he interpreted as foredeeps and suggested that along such 'Pacific-type coasts' the ocean bottom underthrust the continent along the foredeeps. Along the 'Atlantic-type coasts', the continental margin had no relation to the internal structure of the continent. In the Indian Ocean, the western margins were of the Atlantic type, whereas the eastern margins resembled those of the Pacific. The margins of the Antarctic were at that time totally unknown.

Suess' studies since the 1850s had shown him that during the Mesozoic the Mediterranean had reached as far as Central Europe and the Alps had housed deposits representing a much deeper basin than those further north. He and his students and colleagues, such as Ferdinand Stoliczka working in the Himalaya and Melchior Neumayr, later noticed that along the entire Alpine-Himalayan mountain ranges a dominantly Mesozoic seaway existed that in places had depths exceeding 4 000 metres, based on analogies of its sediments with the present-day oceanic deposits. Suess thought this basin to have been an ocean much like the Atlantic and called it Tethys, after the sister and consort of the Greek god of the ocean, Okeanos. South of the Tethys, the distribution of the late Palaeozoic to medial Mesozoic terrestrial deposits and fossils suggested the presence of a vast continent that embraced South America, Africa, India and Australia. Suess called it Gondwana-Land, after the Gondwana deposits in India (in the last volume of the *Antlitz*, Suess considered Australia and Antarctica as an entity distinct from Gondwana-Land). A similar landmass to the north of the Tethys he called Angara-Land. Eurasia had formed by the welding of pieces of Gondwana-Land to those of Angara-Land by the elimination of the intervening Tethys as a consequence of the approach of the two continents. The Atlantic and the Indian oceans were born when parts of Gondwana-Land foundered as the Tethys disappeared. Thus, gradual changes of the older elements of the face of the earth have led to the expression it presently carries.

Suess thought that oceanic subsidences created negative movements of the strand, leading to regressions. He thought that in Iceland he could see ongoing processes of ocean making and this he thought consisted of subsidence and volcanicity. In East Africa he also recognized the Great Rift Valley for the first time as a major tectonic structure and thought its marginal faults and volcanoes similar to those seen along the Atlantic margins, but realized that in East Africa lithospheric stretching clearly was a cause. Suess said this was extension caused by planetary contraction. Nobody could understand what he meant, until in 1987 Sean Solomon from the Massachusetts Institute of Technology showed how this could happen as a consequence of membrane stresses.

Suess' dislike of the uplift theory was such that it led him to deny the ongoing rise of Scandinavia. He also had difficulty fitting isostasy, as gravitational equilibrium between different lithospheric and mantle columns, into his world picture and declared himself a heretic in regard to it as late as 1911. His picture of the interior of the earth was based on Gabriel Auguste Daubrée's ideas derived from the composition of meteorites and Emil Wiechert's seismic studies but did not include isostasy. When the German meteorologist, polar researcher and geophysicist Alfred Wegener and Émile Argand, a Swiss geologist, combined his tectonics with isostasy, the result was continental drift and much of what we have since rediscovered in continental tectonics.

Looking back from today's vantage point, it is incredible how much closer we are to Suess' tectonics than to those of most of his successors in the 20th century before plate tectonics. The face of his earth included tangential and radial movements, the latter consisting entirely of subsidence (which we have partly replaced by intracontinental extension to open oceans). His tectonics was much cited, but little understood (even by modern science historians), because he did not make it easy. He wrote no short and handy textbook as did his less sophisticated successors, such as Leopold Kober, Hans Stille or Walter Hermann Bucher, an American geologist, and careless critics, such as Ferdinand Löwl and Alexander Supan. His *magnum opus*, the *Antlitz*, hid many of his interpretations amidst long and detailed regional descriptions. One had to study the *entire work* carefully to understand its message. The regional plan obscured the theoretical 'long argument' aspect of the book, which was entirely directed against the uplift theory. Those few who understood what he said became mobilists. It proved unfortunate for mobilists, though, that Wegener presented his theory with so little geology and in places with so little attention to detail, so that neither Suess' countryman Otto Ampferer or his son Franz Eduard Suess, nor Argand, nor Alexander Du Toit from South Africa were later taken seriously. The Suessian way of viewing the planet had to slumber until the Canadian geologist and geophysicist J. Tuzo Wilson resurrected it, first in a contractionist framework in the 1950s, but then, in 1965, in the form of plate tectonics. It is plate tectonics that shape the face of the earth, with horizontal motions dominating, much larger than those Suess could possibly imagine.

Water from pristine areas. The Kläffer spring originates in a karstic massif in Styria and delivers 217 million litres of water per day to the Second Vienna Spring Water Main. The photograph shows the overflow of the spring.

Erinnerungen, S. Hirzel, Leipzig, 1916, pp. 153–154

Der Grundsatz, von dem ausgegangen werden mußte, war, daß zum menschlichen Genuße das reinste erreichbare Wasser unter Überwindung aller Schwierigkeiten geboten werden soll. Da die gefährlichste Verunreinigung, nämlich jene organischen Ursprunges, an den Wohnstätten der Menschen haftet, mußten Infiltrationsgebiete gesucht werden, die außerhalb der Besiedelung liegen. Solche waren nur auf den Hochflächen der Alpen vorhanden und die Frage war nun, ob die Quellen an dem Fuße dieser Hochflächen (Schneeberg, Rax, u. a.) trotz ihrer Entfernung in Vorschlag zu bringen seien.

The underlying principle was to provide the purest possible water supply for human consumption by overcoming all potential difficulties. Since the most dangerous pollution is of organic origin and comes from human dwellings, we had to look for infiltration areas situated away from settlements. Such areas are only found at higher elevations in the Alps, which raised the question as to whether springs at the foot of these areas (e.g. Schneeberg, Rax) could be proposed despite the great distance.

TWO WATER PROBLEMS OF A BIG CITY

Günter Blöschl
Vienna University of Technology

Many big cities across the world have been founded on the banks of a river. The reason for this choice of location will often have been the role of the river as a trading route. An additional advantage, depending on individual circumstances, might be water supply, food supply from fishing and sometimes the river as a bastion against enemy attacks. A common disadvantage is the danger of flooding, a problem shared by many big cities.

Another major issue for big cities is water supply. The high population density means that ground and surface waters may become polluted with wastewater, which has for centuries led to problems with drinking water hygiene in many cities and caused epidemics.

For centuries Vienna was the largest city in the German-speaking countries, with dynamic population growth. Around the year 1700, Vienna had about 100 000 inhabitants, one hundred years later, the population had doubled. By 1850, Vienna had half a million inhabitants and just 25 years later it had doubled again. This increase was mainly due to constant immigration from across the Austro-Hungarian monarchy into the capital and also to the high birth rates of the time. The dramatic increase in population brought the two water issues to a head around the middle of the 19th century. Both problems were solved almost simultaneously – by the regulation of the Danube and the construction of the First Vienna Spring Water Main – and in both cases the same key individual, geologist and politician Eduard Suess, pursued far-sighted, sustainable solutions. In this way he shaped the 'face of the earth' in the city of Vienna on these major water issues. His ideas have left their mark to this day.

The urban area of Vienna has been settled since the Stone Age. In the 1st century AD, the Romans adapted the Celtic settlement in the area of today's inner city to create the military post of Vindobona with an adjacent civil town. It was situated on the site of the present city centre, about 15 metres above the level of the river, on a terrace of gravel, roughly 20 metres thick, overlaid with loess. This meant that Vienna was never inundated, even in the highest floods. Northeast of the terrace, in the zone of recent meanders, the Danube branched out into a myriad of larger and smaller arms. These changed their path with each flood event through the interaction of bank erosion and aggradation. In the Middle Ages, the arms immediately next to the city began to silt up and the arms further away carried more and more water. From about 1700, the northernmost arm carried the most water. Wooden bridges, which were swept away in each large flood event, created only temporary links with the north. The population increase in the 17th and 18th centuries, however, meant that Vienna grew in all directions, including northeast towards the Danube on to the lower lying terrain affected by the recurring flood events.

In hydrological terms, the Danube at Vienna has a very heterogeneous catchment area, with two quite diverse main tributaries. The river Inn has an Alpine character, with high precipitation in summer and snow melt in spring. The steep topography means an abrupt runoff of the water from the catchment areas. The second large tributary is the Bavarian Danube, with less precipitation and a flatter topography, but often high soil moisture in winter. Floods can therefore come about through two process causes: summer floods as a result of large-scale intensive precipitation caused by incoming atmospheric humidity from the Mediterranean; and winter floods after less intensive precipitation on to snow cover or saturated ground in lower-lying areas. Vienna also used to struggle with a third type, namely ice jams. During bouts of intense cold, the water of the Danube froze. If temperatures then rose fast and it rained, the blocks of ice were pushed together by the waters of the Danube and jammed in the many branches of the river system. Large-scale flooding and damage to the Viennese suburbs ensued.

As in other European cities, various attempts were made in Vienna to protect the city from flood events. A system of levees was constructed in the second half of the 18th century. Shortly after the work was completed, however, the 'All Saints flood' of 1787 destroyed the levees again. This made it clear that levees could be no long-term solution for Vienna. In the early 19th century, the idea was advanced to move the many branches of the Danube into one joint new bed. In addition to protecting against floods, this measure would benefit shipping and would allow the construction of permanent bridges.

Multiple channels of the Danube in the flood plain near Vienna in the early 19th century.
Detail from *View from the Vienna Woods towards Vienna* by Jakob Alt. Oil on canvas, about 1830.

For decades the responsible expert committees debated various planning variants without coming to an agreement. The rain-on-snow flood of February 1862, which destroyed large parts of the city of Vienna, acted as the immediate trigger to push a speedy solution of the flooding problem to the top of the priority list.

A Danube Regulation Committee was established, with Eduard Suess, as geologist and member of the Lower Austrian provincial assembly, responsible for the technical aspects of the project. The first issue was one of planning the route of the new bed for the Danube, yet to be dug. There were two opposing views: one group favoured a variant to widen the existing, arc-shaped route of the largest Danube arm, which should take up the whole river discharge. The other group favoured digging a completely new bed that would shorten the existing arc, a cut-off on the shortest line. Domestic and foreign experts were called in, but no clear preference for one or the other variant could be established. In the end, Eduard Suess fought for the shortest cut-off. This variant would have the disadvantage of the need for much bigger earthmoving work, but it had the advantage of a more stable bed. Suess wanted to prevent the Danube from relocating its bed again as it had done in previous centuries. The new earth-moving equipment available in the mid-19th century made it possible for the first time to carry out excavation work on such a scale. The Suez Canal, which had necessitated excavation on an even bigger scale, was opened in 1869. At the opening of the Suez Canal, Suess discussed the construction risk of the Viennese plan with the engineers. In the same year, a decision was taken in Vienna to go for the shortest cut-off route. The firm of Castor, Couvreux et Hersent, who had constructed the Suez Canal, were commissioned with the project and moved their equipment from the Suez Canal to Vienna.

Construction started in May 1870. A bed, seven kilometres in length and 280 metres wide, was excavated in the dry. Gustav Wex was the master of the works, supported and advised by Eduard Suess. Together they had to find solutions for further technical problems. Fast inflow of the water into the dry bed had caused great damage in the Suez Canal, and with the large gradient of the Danube, bigger problems might be in store. Suess had the banks of the new bed of the Danube fortified to protect it. When on 15 April 1875 Suess gave the signal for the inflow of the water into the new bed to begin, everything went according to plan and the banks by and large remained stable. The old bed was cut off and most of it filled in. Now the Danube in Vienna consisted of a main bed and a 450 metre wide foreland, destined as overflow area. This was the completion of the Great Vienna Danube Regulation. Permanent bridges began to be built and navigation on the river was ensured.

The main aim of the regulation, however, had been flood protection and the next decades showed that this objective had been fully met. Ice jams hardly caused any damage any more, as the ice blocks jammed only very rarely. The large floodwaters were carried out in the new bed and into the overflow area. The regulation proved particularly beneficial during the flood event of September 1899. Massive precipitation in the catchment area had caused more than 1000 square kilometres of land to be flooded in the catchment area of the Danube. In Vienna itself, the largest flood discharge for more than a century was recorded, but there was little flooding, even though the top water level in some places was just 20 centimetres below the crest of the levees. Smaller flood events no longer presented any problems.

There were no big flood events in the first half of the 20th century. As in many rivers across the world, the flood events at the Danube form temporal clusters – decades with many flood events alternate with those with few flood events. It was not until 1954 that a big flood event occurred. In the meantime, Vienna had grown again and so there was considerable damage.

Like a hundred years earlier, planning started again and went on for years. In the end, the Second Danube Regulation was passed to improve flood protection. The route for the bed favoured by Eduard Suess was retained. Within the overflow area a new, 210 metres wide, bypass channel was created in parallel with the Danube. In the event of a flood, the weirs are opened and the water is discharged into the channel. The bypass channel raised the capacity by half, that means the flood discharge that would not cause overflow was half as much again than before. The maximum discharge is equivalent to the calculated discharge of the flood of 1501, which is generally considered the largest flood of the second millennium.

Recently Vienna has been hit by some big floods again. In August 2002 there was a large flood event and in June 2013 an even bigger one, similar in magnitude to the flood of 1899. In June 2013 the extreme precipitation in the catchment area, together with high soil moisture following a humid spring, caused the large flood discharges. The new flood protection did its job in both cases and there was little damage. Flood protection in Vienna is clearly a success story and Eduard Suess had an essential part in it.

The second big issue in nearly all big cities across the world is one of ensuring water supply for the population and industry. In Vienna, water was for a long time exclusively supplied via domestic wells. The substratum of Vienna lies on different levels of Ice Age terraces covered by several metres of loess and loam. The substratum is mostly made up of fine-grained sediment, bluish-grey when fresh, called *Tegel* in Vienna. Overall, it is a sequence of maritime to brackish to limnic sediments and reflects the gradual drying up of the Vienna Basin. Given the heterogeneous situation, the domestic wells did not yield an abundant supply of water in all cases.

Around the year 1800 some small water supply systems were constructed. In the western parts of Vienna, the impermeable Tegel was unsuitable for domestic wells. In response, the Albertinian Water Pipeline was constructed. It brought water from springs in the Vienna Woods to some western districts. In the centre of Vienna, water supply from the domestic wells had become inadequate. This led to the construction of the Emperor Ferdinand Water Pipeline. On the right-hand bank of the Danube, a trench was dug that ran parallel to the river and was filled with gravel. The water level in the trench was lower than the water level of the Danube, so the water of the Danube seeped into the trench through a 60 metre wide stretch of ground, which filtered the water. This water, from river bank filtration, was piped into the city of Vienna and made available to the population for free through public open wells. Most of the water supply however was still provided through domestic wells.

With the increase in population, the quantity of the water from river bank filtration no longer sufficed. Especially in winter, when the Danube water levels were low, shortages arose. In the summer months the water quality was often inferior because of high temperatures. Cesspits and domestic waste near the filtration site increased the risk to hygiene. More and more domestic wells became polluted. While Vienna city centre, as the first city in Europe, enjoyed a complete sewage system from as early as 1739, the sewers in the suburbs were mostly open ditches, which meant that faeces could seep into the groundwater. Whenever the streams of Vienna flooded, faeces from the sewers were swept directly into the domestic wells. Such pollution caused enormous problems of hygiene and numerous epidemics, particularly typhoid fever and cholera. The cholera epidemic of 1830 caused a death toll of 2000, the typhoid fever epidemic in the summer of 1838, triggered by local flooding of Viennese streams, was also devastating.

In this situation Vienna's water supply problems resembled those of many other big cities in Europe. In London, for instance, wastewater flowed directly into the Thames. This not only caused problems of odour – the summer of 1858 earned the label 'The Great Stink' – but also cholera epidemics, as drinking water was taken from the Thames. An epidemic in 1854 became famous through the records of physician John Snow, who mapped the deaths in the various houses and demonstrated that polluted water had been responsible for the people falling ill. At the same time, Paris experienced numerous epidemics as a result of polluted drinking water.

In Vienna the problems with water supply prompted the municipality to establish an expert committee in 1858, which should assess the situation and suggest improvements for water supply and wastewater treatment. Eduard Suess, when he became professor of geology at the University of Vienna, had published a study on the soils of Vienna. This publication had attracted the interest of the city councillors. He was brought in as geology expert on the issue of water supply and voted onto the city council one month later. In this way Eduard Suess became a key figure in the development of the water supply for Vienna.

The committee essentially discussed two options. The first option was expanding the existing system for river bank filtration from the Danube, increasing the water quantity and improving the filtration. The second option was to make use of water from a large poreous aquifer underlying the populated area south of Vienna, the southern Vienna Basin.

Eduard Suess did not agree with either of these options. He pointed out: "The underlying principle was to provide the purest possible water for human consumption by overcoming any obstacles." This statement reflects his perspective of sustainability and responsibility for the people. Suess saw neither the river bank filtration from the Danube nor the groundwater from the plain south of Vienna as a solution. Instead, he looked for infiltration areas far away from settlements and identified high lying areas in the Alps southwest of Vienna. Suess teamed up with other members of the committee, like engineer Karl Junker, and worked out a proposal for the city council. The report envisaged accessing springs at the foot of the Rax and Schneeberg massifs in the Limestone Alps and piping the water into Vienna via a transfer system about a hundred kilometres in length, with aqueducts and tunnels.

The proposal was seen by many as utopian. There were doubts about the technical feasibility of a water transfer system of such a length and the costs of construction were considered unnecessarily high. Therefore the proposal at first met with strong opposition from the public as well as from the city council. There were, however, also numerous supporters, e.g. deputy mayor Cajetan Felder and the k.k. Gesellschaft der Ärzte (Imperial Society of Physicians) who were all in favour of Suess' principle of supplying the cleanest possible water.

The idea proposed by Suess won out eventually. Vienna city council passed first the planning and then the construction of the water main. Suess explored the hydrogeological situation of the springs at Rax and Schneeberg. Average annual precipitation in the Rax and Schneeberg area is about 1500 millimetres per year, as opposed to 600 millimetres per year in Vienna, so the groundwater recharge is several times that of the Vienna area.

In summer 1866 a cholera epidemic broke out and claimed nearly 3000 lives in Vienna, giving the project additional urgency. The then director of the general hospital wrote an open letter to Suess, in which he reported that by adding large amounts of water to the toilets and sewers of the general hospital each day, it had been possible to massively reduce infections there and that he expected a similar effect for the whole of Vienna from the new water supply.

Construction work on the water mains began in April 1870. Soon after, in May 1870, work began on the flood protection on the Danube.

Several springs were lined and the water piped into spring chambers. The high elevation of the springs allowed the entire transfer system to be run as covered channels using the natural gradient. This meant there was no need for pumping stations, a fact that reduced running costs and increased operational reliability. Thirty aqueducts were built in brickwork following Roman design, the longest one spanning more than 1000 metres. Numerous tunnels were created, the longest measuring nearly 3000 metres. Several water reservoirs were built in high lying suburbs, where the water was collected and stored. A network of cast-iron pipes brought the water into the whole city. Domestic buildings had a faucet in the corridor on each floor, a revolutionary achievement at the time.

In May 1873 the Vienna World Exhibition was opened and celebrated industrial progress. The belief in technological possibilities rose to new heights. However, the crash of the Vienna stock market in the same month, triggered by financial market speculations, took some of the shine off very quickly. In the autumn of the same year, the Emperor Franz Joseph Spring Water Main was opened with pomp and circumstance: at the opening moment the water jet of the Hochstrahlbrunnen fountain in the city shot straight up, a completely new sight at the time.

The water main became a symbol of liberation from water shortage and the threat of epidemics. In an effort to speed up the introduction of water pipes in domestic buildings, the water of the domestic wells was checked for health hazards. If a hazard was found, connection to the new water mains network was officially ordered. This policy meant that by 1883 about 80 percent of Vienna's buildings had been connected to the mains, and five years later this reached 90 percent. The benefits of this action soon showed. While in 1871 more than a thousand people had died of typhoid fever in Vienna, that number fell to a tenth within a few years. Cholera, which had regularly broken out in epidemics claiming thousands of lives, disappeared almost completely.

The population of Vienna continued to grow and in the winter months water shortages reappeared. The only recently completed First Vienna Spring Water Main was therefore expanded by several measures. A few additional springs were added, occasionally surface water was also included, but this soon led to problems of hygiene. So the authorities went back to Eduard Suess' original idea of bringing in spring water from the Alps. Springs further away from Vienna, in the southwest, on the Hochschwab massif, were lined and the water transferred into Vienna across 200 kilometres using the natural gradient, exactly following the project idea of Suess. This project was completed and opened in 1910 as the Second Vienna Spring Water Main. It doubled the supply of clean drinking water.

After the First World War, the population of Vienna shrank and so the water supply continued to be sufficient for several decades. Attention focused on protecting the headwaters which were increasingly threatened by developing tourism. In recent decades additional resources have been explored. One uses groundwater from a porous aquifer in the south of Vienna, the other one water from river bank filtration from the Danube. The First and Second Vienna Spring Water Mains, however, remain the backbone of the water supply.

As a pioneer for realizing the two projects – the Danube Regulation and the Spring Water Main – which were extremely expensive and seen by many as utopian, Eduard Suess has shaped the 'face of the earth' in this area. Both projects remain central to solving these two major water problems for the capital city of Austria to this day.

Judd, J. W.: Prof. Eduard Suess, For. Mem. R.S., Nature, 93, 1914, p. 245

The writer of this notice recalls with pleasure the happy time he spent with Suess forty years ago, when he had the opportunity of witnessing the delightful relations that existed between the professor and his students. Not only during geological excursions in the neighbourhood of Vienna was the charm of Suess's society felt, but in the Wurstel-Prater, where we joined the young fellows during hours of relaxation – in the beer-gardens, and even on the "merry-go-rounds." Yet, amid all the fun and frolic, the signs of affectionate respect and devotion to the great teacher were never for a moment wanting.

Viennese lifestyle. The giant wheel in the Wurstel Prater, a traditional gathering place in the city.

MILESTONES OF A LIFE BEYOND THE GEOSCIENCES

Thomas Hofmann
Geological Survey of Austria

"I was a very bad boy and my father was very strict," Eduard Suess recalls in his memories of his childhood. He was born on a Saturday, 20 August 1831, in London, 4 Duncan Terrace, as Eduard Carl Adolph Suess.

The cholera was raging on the continent at the time. It had spread steadily from Russia westwards and become a nightmare. People everywhere felt paralysed and lacked ideas of how to handle this situation. In the respected *Wiener Zeitung*, the local book shops advertised various writings on how to deal with cholera, what prevention measures to take, etc.. Frédéric Chopin, aged 21, wrote on 16 July 1831 of his concert tour from Vienna, "Everybody here is terrified of the cholera. They take their anxiety to ridiculous extremes and are selling printed prayers that implore God and all saints to make the cholera die down. Nobody dares to eat fruit and most are fleeing the city."

Seventy years later many daily newspapers reported Eduard Suess' farewell lecture on 13 July 1901. Many quoted the professor who had begun teaching at the university in 1857 and continued for 88 semesters. The *Wiener Zeitung* quoted Suess on 14 July 1901: "In the course of the 44 years of my activity, much has happened but nothing as radical as the progress made in the natural sciences. They have changed the social, societal and all other circumstances, but if you look closely you will find that alongside the value of natural science, the natural scientist is gaining in significance."

What Suess modestly left out was the fact that he himself had decisively contributed to the progress of the natural sciences, especially the geosciences, during these decades. Moreover, in infrastructure and the transport system, in urban planning and architecture, these decades were so defining that they shape everyday life to this day. And here too, Suess' activities have left their mark. The First Vienna Spring Water Main (Hochquellenwasserleitung), which he initiated and which was opened on 24 October 1873, still supplies 40 percent of the water needed in Vienna in the 21st century.

The life of Eduard Suess is unique, even 100 years after his death. When he was barely 26 years old, on 10 August 1857, Emperor Franz Joseph I. made him an associate professor of palaeontology at the University of Vienna, even though Suess had neither a doctorate nor the lecturing qualification (Habilitation). In 1867 he became a full professor of geology and in 1888/1889 rector of the Alma Mater Rudolphina Vindobonensis, founded in 1365.

His career at the Academy of Sciences developed in parallel. In 1860, aged 29, he became a corresponding member and in 1867, the year when he became a full professor, the Academy accepted him as a full member. From 1885 he was Secretary of the Section for Mathematics and the Natural Sciences, and from 1891 General Secretary. In 1893 he was elected Vice-President. His presidency from 1898 until 1911 was one of the longest periods anyone has steered the scholarly association. In this function he strove for cooperation of the Academy of Sciences in Vienna with those of Munich, Leipzig, Göttingen and Prussia, as well as for an international association of European and US academies. His presidency saw the creation of the first Phonogram Archive in Europe in 1899 and the establishment of the Institute for Radium Research in 1909.

As if this was not enough, geologist Suess – "At the core of his being he was a scientist," said his son Franz Eduard on the 100th anniversary of his father's birthday – also was a liberal politician. In this role he was particularly committed to schools and education. In 1863 he gained a seat on Vienna's municipal council and from 1869 he was active in the Lower Austrian provincial assembly, including as inspector of schools. In 1873 he renounced his seat on the municipal council because he was elected into the Reichstag with 631 votes for and 59 votes against him. He was sworn in on 4 November 1873 and stayed until 1897, another advance in his political career.

The question remains, how could a person that one might today call a workaholic manage so much? Suess had a very strict daily schedule. For many years he lived in Vienna's 2nd District, in Afrikanergasse 9, near the Prater park, where he died on 26 April 1914. His biographer, Vladimir A. Obruchev, writes, "Suess combined an incredible enthusiasm for his work with the ability to make strict use of his time. He got up at half past six, and an hour later, always on foot, he went off to the university where he held his lectures from eight to nine. Even in periods of strenuous political activity he never took research leave or moved lectures.

The morning hours he used for his work for the Academy, the Reichstag or the provincial assembly. After a nap, Suess would walk for an hour in the riverine woodlands of the Prater near his flat and later he used the evening, often until the early hours of the morning, for his research or to prepare talks and lectures, unless he had to attend a meeting of a scientific society, the Reichstag or the provincial assembly." His colleague Theodor Fuchs also described his admiration for Suess' working style on the occasion of his 75th birthday: "Suess' capacity for work was amazing, he was a sheer fanatic of work and activity. It was impossible for him to be inactive. Work seemed to be no effort but fun, tiring was an alien concept for him and his recreation consisted solely of switching from one task to another."

In 1855, Suess married Hermine (née Strauß, niece of Paul Partsch [1791–1856], head of the k.k. Mineralogische Hof-Cabinet and Suess' boss). They had seven children. "To his children, [...], he was an ideal father", recalls his youngest son Erhard (1871–1937). His brothers were Adolf (1859–1916), Hermann (1864–1920), Otto (1869–1941) and Franz Eduard (1867–1941) – the latter became a well-known geologist. His sister Paula Aloisia (1861–1921) married palaeontologist Melchior Neumayr (1845–1890), his sister Sabine (1859–1868) died in childhood.

If one looks for answers to the question 'How could one man be so successful?', the answer is not just diligence and discipline, but also the support of his family.

His parents had married in 1828, in 1829 Louise was born, Eduard in 1831, followed in 1833 by Friedrich. In late autumn 1834 they moved from London where his father had been a wool merchant, via Rotterdam to Prague to join the maternal grandparents. Eduard Suess recalls, "We were quite English children and knew no word of German." Even though German was spoken in the new home, where the youngest child, Emil, was born, the children continued to receive teaching in English. "Our parents, who perceived the advantage that knowing this language would give us one day, procured an English nanny, Miss Gretten." In addition, when Eduard was around five years old, his father brought an English teacher, Mr Augustus Thurgar, from Norwich to Prague to take on Eduard's education. "Teddy boy, do you want to become a gentleman?" "Oh yes, I said, without knowing what he meant by the term 'gentleman', but I agreed because I could see quite clearly that he meant well." When he was seven, he started studying with a private tutor, a theology student nearing the end of his studies, "to learn German well". French he first learned from a mademoiselle, later from an old Belgian described by Suess as, "what was left of the Grande Armée". In 1840 he received a special dispensation to attend the grammar school in Prague, which was German at the time. The head of school, an excellent scholar of Slavic languages, also taught the boys Czech.

If we try to trace Suess' successes, the answers lie in the early support for language learning by far-sighted parents and experienced teachers who were capable of instilling values in the young boy.

When the Suess family moved to Vienna in autumn 1845, the Biedermeier-type intellectual world of 1831 had changed little. Still, some men, notably Wilhelm Haidinger (1795–1871), had pushed the natural sciences forward. He founded the association Freunde der Naturwissenschaften in 1845 and headed both the k.k. Montanistisches Museum, created in 1835, and its successor, the k.k. Geologische Reichsanstalt, inaugurated on 15 November 1849, nowadays the Geological Survey.

Aged 16, Suess entered the Polytechnikum, today's Technical University, and both witnessed and participated in the March Revolution of 1848. "We were fired up by an indescribable feeling, of freedom and patriotism, of enthusiasm and death-defying courage, ..."

In terms of his career, Suess was, as one would say today, 'in the right place at the right time'. After the revolution and the end of the Metternich era, an enormous surge of development started in liberated Vienna. Young talents like Suess were able to succeed quickly in this early vacuum. Institutions like the k.k. Mineralogische Hof-Cabinet, precursor of the Natural History Museum Vienna, where Suess was sworn in as an assistant on 15 May 1852 and where he worked until 1862, or the k.k. Geologische Reichsanstalt, formed the institutional framework, long before discipline-specific university institutes came about, although there Suess' name would crop up again.

For Suess, who initially specialized in working on brachiopods and laid down essential foundations for stratigraphy, fieldwork and travel were always important for expanding his knowledge and his horizon. In 1854 he and Franz von Hauer, 'Premier Geologists' at the k.k. Geologische Reichsanstalt, visited Switzerland together. On their return journey, Arnold Escher von der Linth accompanied them. "Two years later, in summer 1856, a meeting of German natural scientists and doctors took place in Vienna. Merian and Escher really did come to Vienna and I presented the programme of an Alpine Geological Society to the discipline. [...] This society was to span 'the entire spine of Europe, from Lyon to Vienna', as it said in the programme," writes Suess on the occasion of the 50th anniversary of the Austrian Alpine Club in 1912. Even though this idea of the 25-year old Suess did not gain the support of the authorities of the monarchy at the time and his plan folded, we can already discern Suess, the visionary, with his far-flung ideas and concepts that stretched across countries and borders that were often ahead of his time.

In 1862 his first major piece of writing was published: *Der Boden der Stadt Wien* (The Soil of the City of Vienna), with the telling subtitle 'after its formation, texture and its relation to civic life'. Suess himself, with typical modesty, called it a 'geological study'. Today the 326-page volume, which includes a geological map, must be understood as the beginning of 'urban geology', a discipline of the geosciences that has risen in significance in the 21st century.

Suess thought in an interdisciplinary mode. Having observed that the corpse water from higher lying cemeteries seeped into the domestic wells below, he demanded, "It must be prevented at all costs that the groundwater of the cemeteries enters beneath our suburbs...", and called for the abandonment of inner-city cemeteries. Mayor Andreas Zelinka promptly installed Suess in the municipal water supply committee. Shortly after he was also voted onto the municipal council of Vienna. Suess entered the history books with the opening of the First Vienna Spring Water Main on 24 October 1873.

Suess, the geologist, acted on a scientific level, while Suess, the politician, fought for the realization of his ideas on the practical front. As a brilliant speaker he relied on well-argued information for the decision makers. In 1862 he published a series of articles on the Vienna water supply in the weeklies *Wiener Medizinische Wochenschrift*, and in the *Wochenschrift für Wissenschaft, Kunst und öffentliches Leben* – a supplement of the *Wiener Zeitung*.

Suess' success is due to this double function, which characterizes him as a man of action. His son Franz Eduard summed it up concisely in 1931, "Eduard Suess was one of the most significant leaders of the strong party of the German-liberal left, in permanent opposition to the conservative stance of Earl Taaffe's government, and one of the most brilliant speakers of the house. His real profession, however, was that of a geology professor at the University of Vienna."

One of Suess' hitherto un(der)appreciated activities is that of knowledge transfer. On 15 January 1860, the association Verein zur Verbreitung naturwissenschaftlicher Kenntnisse, which still exists today, held its constitutive meeting in Vienna. Its first president was Eduard Suess. He himself gave talks to interested non-experts on various topics, be it loess, the concept of time in geology, the dust in Vienna or the (geological) structure of Europe.

The titles of writings like *Bemerkungen über den naturgeschichtlichen Unterricht an unseren Gymnasien* (Comments on the natural science teaching in our grammar schools) tell of his commitment to education. Suess demanded that teachers should have a solid natural science training, as well as the reintroduction of natural-science subjects in the final exams (Matura or A-level equivalent). Suess also pursued the establishment of a mining academy in Vienna like the one that exists in Leoben. He received no support from the ministers for this plan nor for his demand, put forward together with Friedrich Brauer in 1878, for creating a separate mathematical / natural-science faculty at the University of Vienna. The natural-science faculty did not split off from the philosophical faculty until 1975 (!), so Suess was ahead of his time by nearly a hundred years.

After *Der Boden der Stadt Wien*, Suess completed his second major work, *Die Entstehung der Alpen* (The Origin of the Alps), in 1875. Here Suess not only mentions lateral forces at work in the building of mountains, but goes one step further and distinguishes three layers in the formation of the earth, which he describes thus: "The first is the atmosphere, the second is the hydrosphere, the third is the lithosphere." Starting from this tripartite division, he concludes, "and on the surface of the solid [layer] we can distinguish an independent biosphere." Such sentences show Suess' intellectual world. Starting from details, he developed his ideas ever further, opening up new dimensions and perspectives. The published concepts make up the foundation to his *magnum opus*, *Das Antlitz der Erde* (The Face of the Earth), the first volume of which came out in 1883. The second part of the third volume was published in 1909; translations into English, French, Italian and Spanish make this work a milestone for geology to this day.

Suess, who in his numerous leading functions was a public figure for decades, did not just have his admirers and students who hung on his every word – like Emil Tietze, "It was always a joy to listen to Suess, whether you agreed with his position or not, ..." – but also his critics. One of those was Vienna district mayor Konrad Ley, who strictly opposed the First Vienna Spring Water Main project, "We older ones just had the Danube and look what strong lads we've grown into, and now we are supposed to waste millions. Our Mr Suess is a nice gentleman, but a professor, after all." Fifteen years after the opening of the water supply line, the number of deaths from typhoid fever had fallen dramatically, from previously 34.21 per 1000 inhabitants to now 9.44 per 1000. Suess summed it up succinctly, "The main aim of the work has thus been achieved. The merit of the geologist in this was secondary."

Some contemporaries found that Suess was more of a journalist than a researcher and his scientific writings closer to feuilletons. This is the origin of the label 'Geo-Poet' for the great scientist. Theodor Fuchs saw it as a compliment and commented on it on the eve of Suess' 75th birthday: "... he wanted to get directly involved in the vibrant stream of scientific, intellectual life – in this sense he was a journalist. Suess' writings are often short, they contain no superfluous additions, they are pleasantly expressed and in this sense they are feuilletons. But nearly every one of these feuilletons has had a trailblazing impact, became the starting point for a new direction of research. And now even a 'Geo-Poet'!"

Suess, whose life may encourage the emergence of myths, had far-sighted ideas and visions. He put his expert knowledge to good use, in combination with his networking and his gift for captivating an audience. Eduard Suess, whose name has a place of honour in all well-known scientific associations of his discipline, showed not just academic but also human greatness. Franz Eduard recalls, "His whole being was the most natural and at the same time most noble simplicity, utterly free from overambition or vanity, envy or contempt for others. With continued steadfastness he rejected any distinction from above."

NB.: If the human impact on the atmo-, bio, hydro- and lithosphere has increased dramatically, leaving traces on the face of the earth, then the life of Eduard Suess suggests solutions. It needs great, far-sighted and yet modest figures, who, having enjoyed a comprehensive theoretical and practical education, possess the courage, not only to make the right decisions for the future but also to implement them.

Educating the public. The most intense and convincing knowledge transfer in earth system sciences happens in nature's classroom. Biodiversity Day, Wienerwald Biosphere Reserve.

Ueber die Entstehung und die Aufgabe des Vereins, Schriften des Vereines zur Verbreitung Naturwissenschaftlicher Kenntnisse in Wien, 1, 1862, pp. 9–10

Wir vermögen nicht, Ihnen an Winterabenden den unmittelbaren Naturgenuss einer schönen Landschaft herzuzaubern, aber wir nehmen die einzelnen Theile aus dem Bilde und lehren Sie dieselben besser zu betrachten. Der Bau des Gebirges, auf welchem Sie gestanden, die Organisation der Pflanzen, die Sie auf demselben trafen, selbst die Luftströmungen, die Sie empfanden, ja sogar die Natur der erleuchtenden Sonne, solches sind die Gegenstände unserer Vorträge und wenn Sie nach diesen im Sommer wieder hinaustreten in die offene Natur, dann hat sich, so hoffen wir, zu Ihrer früheren Freude auch ein etwas höherer Grad von Verständniss gesellt, Sie wissen der Natur tiefer in ihr grünes Auge zu schauen und die grössere Innigkeit Ihres Entzückens lehrt Sie, wie schön der Beruf des Naturforschers sei.

On winter evenings we cannot conjure up the immediate natural delight of a beautiful landscape, but we take individual parts from this image and teach you how to observe them better. The formation of the mountains on which you stood, the organization of the plants you encountered there, the flows of air you experienced; indeed, even the nature of the illuminating sun. These are the subjects of our lectures and, when you next step out into open nature in the following summer, we sincerely hope that to your earlier joy a somewhat enhanced degree of understanding is added. You will know how to look deeper into nature's green eye and the greater intimacy of your delight will teach you the beauty of the natural scientist's profession.

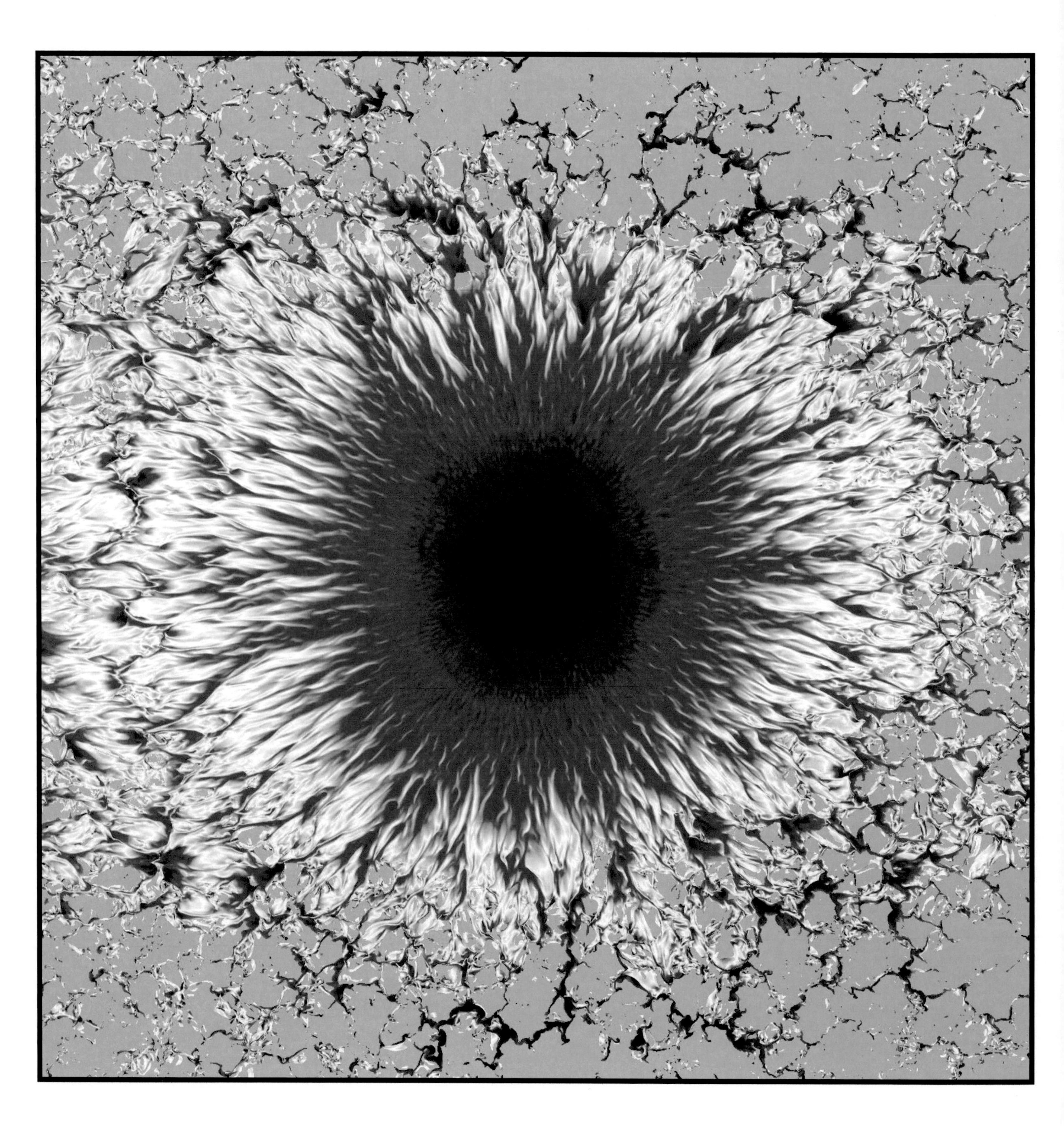

Scientific innovation. Simulated eruption from the surface of the sun producing a phenomenon known as sunspots.

Ueber den Aufbau der mitteleuropäischen Hochgebirge, Anzeiger der Kaiserlichen Akademie der Wissenschaften, Mathematisch-Naturwissenschaftliche Classe, Wien, 10, 1873, p. 131

Der Verfasser gelangt zu dem Schlusse, dass die gesammte Erdoberfläche sich thatsächlich in einer allgemeinen, aber überaus langsamen und ungleichförmigen Bewegung befindet, welche in Europa zwischen dem 40. und 50. Breitengrade gegen Nordost oder Nord-Nordost gerichtet ist. Die sogenannten alten Gebirgsmassen bewegen sich dabei langsamer als die zwischen ihnen liegenden Regionen, welche Ketten bilden, die sich aufstauen und in welchen in Mittel-Europa an der polaren Seite regelmässige Falten, an der aequatorialen aber Risse erzeugt werden. Diese eigene Bewegung der Erdoberfläche verhält sich zur Bewegung der ganzen Planeten etwa so, wie die sogenannte eigene Bewegung der Sonnenflecken zur Rotation des gesammten Sonnenkörpers und ihre Richtung ist in verschiedenen Theilen der Erdoberfläche eine verschiedene.

The author comes to the conclusion that the entire surface of the earth is indeed in motion, a general but very slow and non-uniform movement, which, in Europe, is directed to the northeast or north-northeast between the 40th and 50th degree of latitude. The so-called old groundmasses move more slowly than the regions between them, forming chains that pile up and where, in Middle Europe, continuous folds develop on the polar side and rifts on the equatorial side. This movement of the earth surface relates to the movement of the entire planet in a similar way as the so-called proper motion of sunspots relates to the rotation of the sun, and it takes different directions in different parts of the earth.

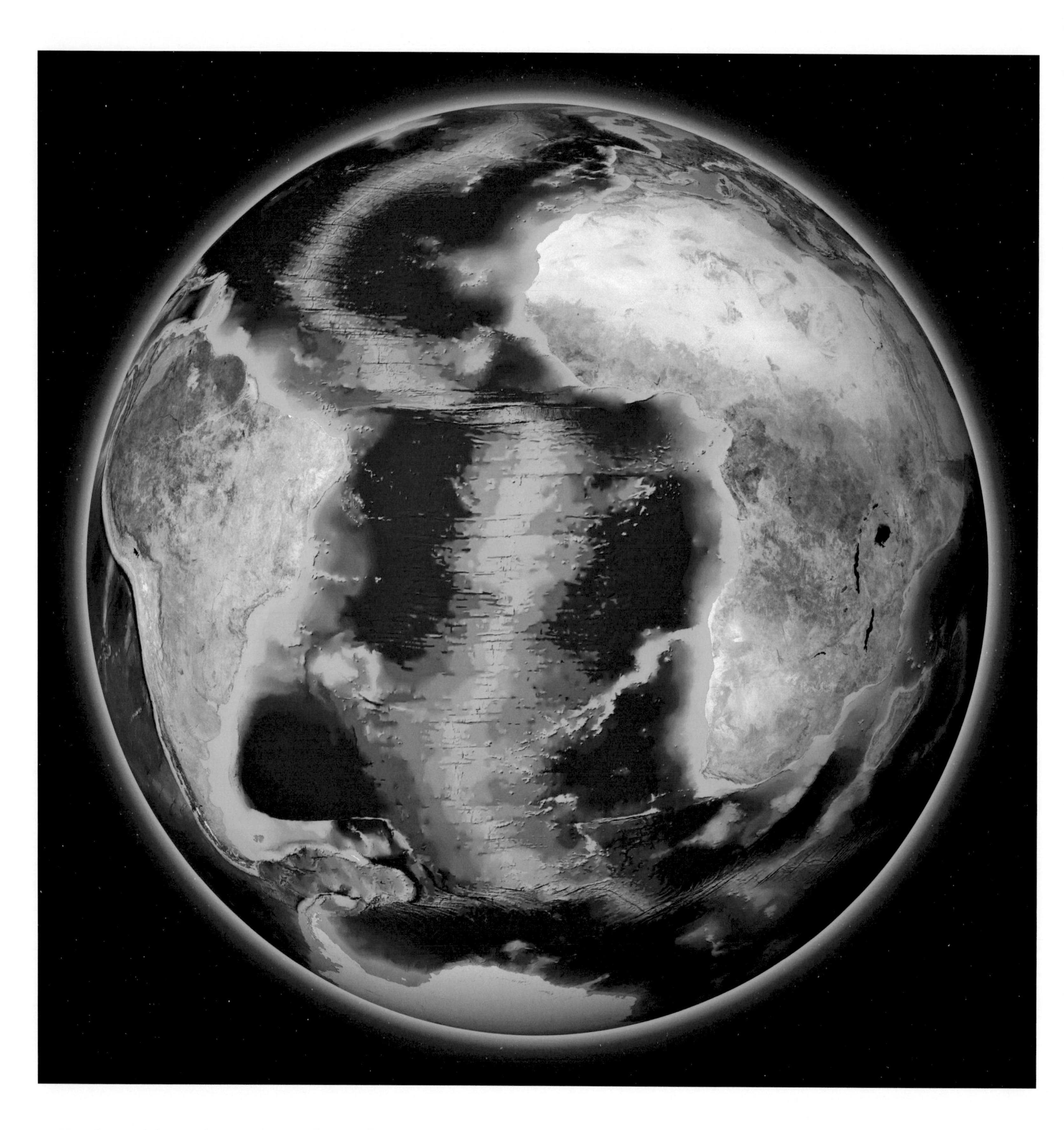

The shape of the continents. The matching outlines of South America and Africa (visualized by a composite satellite image) result from the breakup of the supercontinent Gondwana.

The Face of the Earth Vol. I., Clarendon Press, Oxford, 1904, p. 1

If we imagine an observer to approach our planet from outer space, and, pushing aside the belts of red-brown clouds which obscure our atmosphere, to gaze for a whole day on the surface of the earth as it rotates beneath him, the feature beyond all others most likely to arrest his attention would be the wedge-like outlines of the continents as they narrow away to the South. This is indeed the most striking character presented by our map of the world, and has been so regarded ever since the chief features of our planet have become known to us. It recurs in the most diverse latitudes: Cape Horn, the Cape of Good Hope, Cape Comorin in the East Indies, Cape Farewell in Greenland are some of the best known examples.

Über die Brachiopoden der Kössener Schichten, Denkschriften der Kaiserlichen Akademie der Wissenschaften, Mathematisch-Naturwissenschaftliche Classe, 7, 1854, p. 29

Der Zweck der Abhandlung, welche ich hiermit der Öffentlichkeit übergebe, ist, durch die Untersuchung der Art und Weise, wie eine specielle Thiergruppe in einer abgegrenzten Stufe unseres Hochgebirges vorkömmt, auf der einen Seite einen Beitrag zur paläontologischen Vergleichung und Parallelisirung dieser Schichten mit fremden Vorkommnissen zu liefern, auf der anderen Seite aber, vom zoologischen Standpunkte ausgehend, die Eigenthümlichkeiten der Formen zu zeigen, welche diese Thiergruppe bei uns bietet.

The purpose of the essay investigating the manner in which a specific group of animals occurs in a distinct altitudinal zone of our high mountains, which I thus submit to the public, is to contribute, on the one hand, to the palaeontological comparison and parallelization of these layers with incidents elsewhere and, on the other, to highlight, from a zoological point of view, the peculiarities of this group of animals in our region.

Remnants of an ancient biosphere. Fossil collections are the base of palaeobiological and stratigraphic research. Brachiopods from the collection of the Geological Survey of Austria, which Suess himself once studied.

Urban water. The fountain at the Schwarzenbergplatz in Vienna, symbolizing the start of a new era of centralized water supply. The new water supply system led to a dramatic drop in fatalities from typhoid fever and cholera.

Erinnerungen, S. Hirzel, Leipzig, 1916, pp. 242–243

Vier Tage später, am 24. Oktober [1873], wurde die Wasserleitung eröffnet. Viele Tausende von Wienern füllten den weiten Schwarzenberg-Platz und die benachbarten Straßen. Felder ersuchte großmütig mich, der ich doch das Werk verlassen hatte, durch Erheben eines weißen Tuches das Zeichen zum Öffnen des Wechsels zu geben. [...] Ich stand neben dem Bürgermeister und der Gemeindevertretung am Rande der Terrasse vor dem Schwarzenbergschen Palaste; in unserer unmittelbaren Nähe befand sich der Kaiser. „Per tot discrimina..." dachte ich bei mir, und gab das Zeichen mit dem Tuche. Die Augen der Menge sind auf die Mitte des Wasserbeckens gerichtet. Es ist nichts. Eine peinliche Pause. Nach einigen Minuten wiederhole ich das Zeichen. Wieder nichts. Eine noch peinlichere Pause. Eine, zwei, drei Minuten. Ich beginne die Pulse an meinen Schläfen zu verspüren. Eben, indem ich im Begriffe bin, ein drittes Zeichen zu geben, zeigt sich ein Aufsprudeln an der Mündung des Steigrohres. Höher und immer höher erhebt sich in schwankendem Spiel der schaumweiße Strahl, erreicht endlich, alle Häuser übersteigend, 40 bis 50 Meter und löst sich in eine Spreu von Millionen herabfallender Tropfen auf. In diese sendet die gütige Sonne ihre Strahlen und spannt einen Regenbogen um das Bild. Ein vieltausendstimmiger Ruf des Staunens füllte den weiten Raum. Mir schnürte sich die Kehle zusammen. Mein Blick suchte in der Menge meine gute Frau; ich fand sie nicht. Dann führte mich Felder zum Kaiser. Nach äußerst gütigen Worten der Anerkennung sagte der Kaiser: „Ich danke Ihnen." Ich gab meiner Freude darüber Ausdruck, daß dieses Werk angewandter Naturforschung unter der Regierung Sr. Majestät zustande gekommen sei.

Four days later, on 24 October [1873], the water main was opened. Many thousands of Viennese filed into Schwarzenbergplatz and the adjoining streets. Although I had already left the project, Felder generously asked me to give the signal for opening the change cock by raising a white flag. [...] I stood next to the Mayor and the municipal representatives on the edge of the terrace in front of the Schwarzenberg Palace; the Emperor was very near us. "Per tot discrimina...," I thought to myself and gave the sign. The eyes of the crowd focused on the centre of the basin. Nothing happened. An embarrassing pause. After a couple of minutes I repeated the signal. Again nothing. An even more embarrassing pause. One, two, three minutes. I started to feel my temples throbbing. Just as I was about to give a third signal, there was a bubbling at the opening of the standpipe. Higher and higher, playfully swaying, the foamy white gush of water rose, finally reaching 40 or 50 metres in height, higher than all the buildings, and dissolving into millions of falling droplets. Into these the benevolent sun sent its rays and spanned a rainbow over the entire scene. A cry of astonishment from many thousands of voices filled the wide space. I had a lump in my throat. My eyes searched the crowd for my dear wife; I could not see her. Then Felder led me to the Emperor. After most gracious words of appreciation the Emperor said: "I thank you". I expressed my delight that this project of applied natural sciences had been achieved under His Majesty's reign.

Long distance water transfer. The Leobersdorf Aqueduct of the First Vienna Spring Water Main is the longest of its 30 aqueducts. Without the need for pumps, the water runs along the natural gradient from the Alps to the city of Vienna.

Erinnerungen, S. Hirzel, Leipzig, 1916, p. 244

Als die neue Wasserleitung in 91% der Häuser eingeführt war, schätzte der Obersanitätsrat Prof. Drasche die bis dahin erzielte gesamte Verminderung der Todesfälle an Typhus auf 7 961. Ihre Zahl vor 1867–73 [betrug] 34,21 in 1 000 Todesfällen und im gleichen Zeitraum bis 1888 nur 9,44. Dabei war die Einleitung des Wassers nur nach und nach erfolgt. Das Hauptziel des Werkes war daher erreicht.

When 91% of the houses had been connected to the new water main, the chief medical consultant, Prof. Drasche, estimated that a fall in typhoid fever casualties to 7 961 cases had been achieved. In the time before 1867–73, there had been 34.21 deaths from typhoid fever in every 1 000 deaths, but in the period to 1888 only 9.44 per 1 000, even though the connection to the water main had been carried out gradually. The main objective of the project had thus been achieved.

Vienna tames its river. Danube flood waters in June 2013 at the inflow weir into the bypass channel near Langenzersdorf. The channel increases the discharge capacity of the system to 14 000 cubic metres per second.

Erinnerungen, S. Hirzel, Leipzig, 1916, p. 264

Ein Zufall wollte, daß eine der leitenden Personen der Kommission am entscheidenden Tage unwohl wurde, eine andere aber sonstwie verhindert war, so daß mir die verantwortungsvolle Ehre zufiel, am Roller den entscheidenden Befehl zum Einlasse des Stromes zu geben. Das geschah am 15. April um 3.30 Uhr nachmittags. [...] Die dahinfegende riesige Wassermasse spülte bald über den Landstreifen an der Reichsbrückenlinie. Um 7.20 abends öffnete ihn Fänner, so daß nunmehr ein einziges, zusammenhängendes Bett vorhanden war.

By coincidence, a leading committee member was unwell on that special day and another was deterred for some other reason, so that the responsible honour of giving the key order at the dam to let in the river fell to me. This occurred on 15 April at 3:30 pm. [...] Soon the enormous gushing waters flowed over the little dam at the Reichsbrücke. At 7:20 pm Fänner opened it so that now only one single connected river bed existed.

Long distance correlations. Fossil molluscs at the type locality of the Sarmatian Stage, Nexing, Lower Austria, are similar to those from the same time interval in places as far away as the Aral Sea basin.

II. Über die Bedeutung der sogenannten „brackischen Stufe" oder der „Cerithienschichten", Sitzungsberichte der Kaiserlichen Akademie der Wissenschaften, Mathematisch-Naturwissenschaftliche Classe, Wien, I. Abtheilung, 54, 1866, p. 232

Wenn man also diese Stufe als die brackische bezeichnet, so ist der Ausdruck in soferne richtig, als keine andere Abtheilung unserer tertiären Bildungen eine annähernd eben so große Masse an brackischen Einlagerungen umfaßt. Sobald man aber versucht, Vergleiche mit außerhalb dieser Niederung liegenden Bildungen anzustellen, muß dieser auf locale Erscheinungen gegründete Name verschwinden, da, wie wir bald sehen werden, für die weit ausgedehnten östlichen Aquivalente derselben keineswegs die Anzeichen brackischer Bildung vorliegen. Um nun einen solchen Gesammtnamen zu besitzen, werde ich künftighin im Einverständnisse mit dem, um die Kenntniß der östlichen Fortsetzungen so verdienten Herrn Barbot de Marny, diese gesammten Ablagerungen, nämlich unsere Cerithienschichten sammt dem Hernalser Tegel, als die „sarmatische Stufe" bezeichnen, und jene östliche Fauna, zu welcher *Mactra podolica, Donax lucida* u. s. f. gehören, die sarmatische Fauna nennen.

If this stage is referred to as Brackish, this expression is correct insofar as no other branch of our Tertiary formations includes anywhere near as large a mass of brackish enclosures. However, as soon as one tries to compare these formations with those lying outside these lowlands, the term Brackish, based on local phenomena, is no longer appropriate, because, as we shall soon see, there are no signs of brackish formations in its far-reaching eastern equivalents. In order to obtain a common name, I shall therefore, in agreement with Mr Barbot de Marny who has acquired such knowledge of the eastern continuations, in future call these deposits (i.e. our Cerithian layers, including the 'Hernalser Tegel') the Sarmatian stage and shall refer to that eastern fauna, which includes *Mactra podolica, Donax lucida* and many more, as the Sarmatian fauna.

The Face of the Earth Vol. II., Clarendon Press, Oxford, 1906, pp. 537 – 538

T*he crust of the earth gives way and falls in; the sea follows it.* But while the subsidences of the crust are local events, the subsidence of the sea extends over the whole submerged surface of the planet. It brings about a general negative movement. As a first step towards an exact study of phenomena of this kind, we must commence by separating from the various other changes which affect the level of the strand, those which take place at an approximately equal height, whether in a positive or negative direction, over the whole globe; this group we will distinguish as *eustatic movements*.

Unstable sea level. The İyisu marine terraces west of Gelibolu in the Dardanelles Straits, Turkey. They formed in an interval between 209 and 186 thousand years ago by land uplift combined with a higher global sea level.

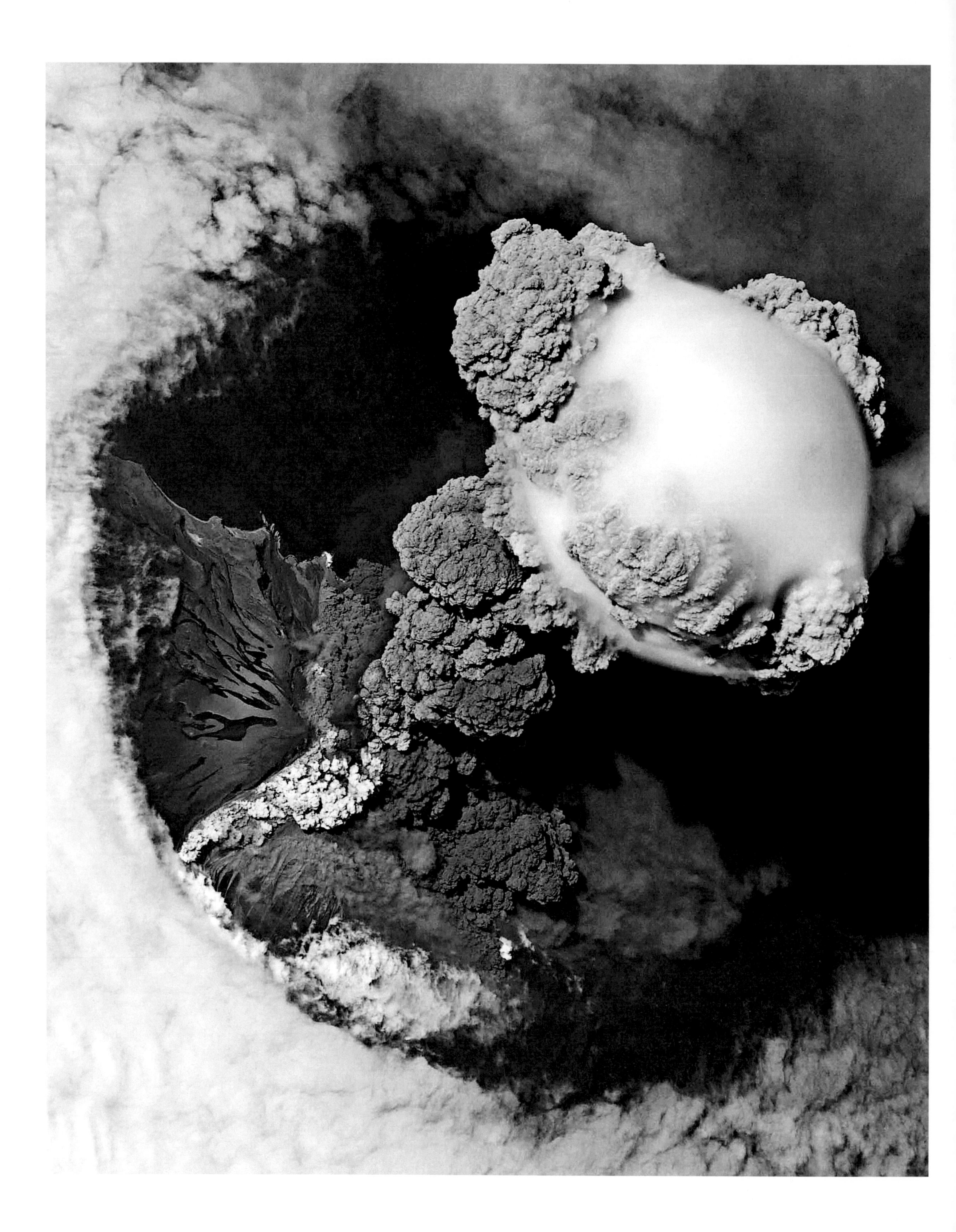

The Face of the Earth Vol. IV., Clarendon Press, Oxford, 1909, p. 548

The recognition of this fact led to the distinction of two kinds of waters. The *vadose* waters – a name originally chosen by Pošepny for the waters which infiltrate from the surface and escape from mineral lodes – include all the waters of the earth's surface, such as oceans, rivers, clouds, atmospheric precipitations, and artesian springs. *Juvenile* waters, on the other hand, are those which arise when the hydrogen issuing from the earth's interior, under very high pressure and at a very high temperature, combines with the oxygen of the atmosphere. The white balls of steam emitted by the volcano become clouds, and a juvenile rain pours down its slopes. The juvenile hot springs bring up unexpected mineral matters from the depths.

Two kinds of water. Juvenile water vapour and volcanic ash ejected in a volcanic eruption on 12 June 2009 at Sarychev Peak, Kuril Islands.

Are Great Ocean Depths Permanent? Natural Science, 2, London, 1893, p. 183

Modern geology permits us to follow the first outlines of the history of a great ocean which once stretched across part of Eurasia. The folded and crumpled deposits of this ocean stand forth to heaven in Thibet, Himalaya, and the Alps. This ocean we designate by the name 'Tethys', after the sister and consort of Oceanus. The latest successor of the Tethyan Sea is the present Mediterranean.

The lost ocean: Tethys. Various stages in the evolution of the Tethys Ocean, from the Late Triassic (210 million years ago), Late Jurassic (150 million years ago) to the Early (120 million years ago) and Late Cretaceous (90 million years ago) (left to right).

The rise and fall of continents: Gondwana-Land. *Glossopteris* leaves, in the collection of the Natural History Museum Vienna, as evidence for the existence of the southern continent Gondwana.

The Face of the Earth Vol. I., Clarendon Press, Oxford, 1904, pp. 595 – 596

If we attempt to make similar comparisons with regard to the united mass of Asia, Africa, and Europe, it at once becomes evident that heterogeneous regions – the limits of which do not coincide with the recognized boundaries of these subdivisions – have here been welded together to form one great continent. The first region comprises the southern and a great deal of the more central part of Africa, then Madagascar and the Indian peninsula. The lofty table-lands of this region have never, so far as we know, been covered by the sea since primitive times, or the end of the Carboniferous period; it is only at the foot of the table-lands that marine sediments have been deposited, which followed the encroachment of the Indian Ocean, as this was formed by subsidence within the tabular mass. We call this mass Gondwána-Land, after the ancient Gondwána flora which is common to all its parts; it corresponds to a large extent with the Lemuria of zoologists; judged from the standpoint we took before, this country is incomparably older than North America.

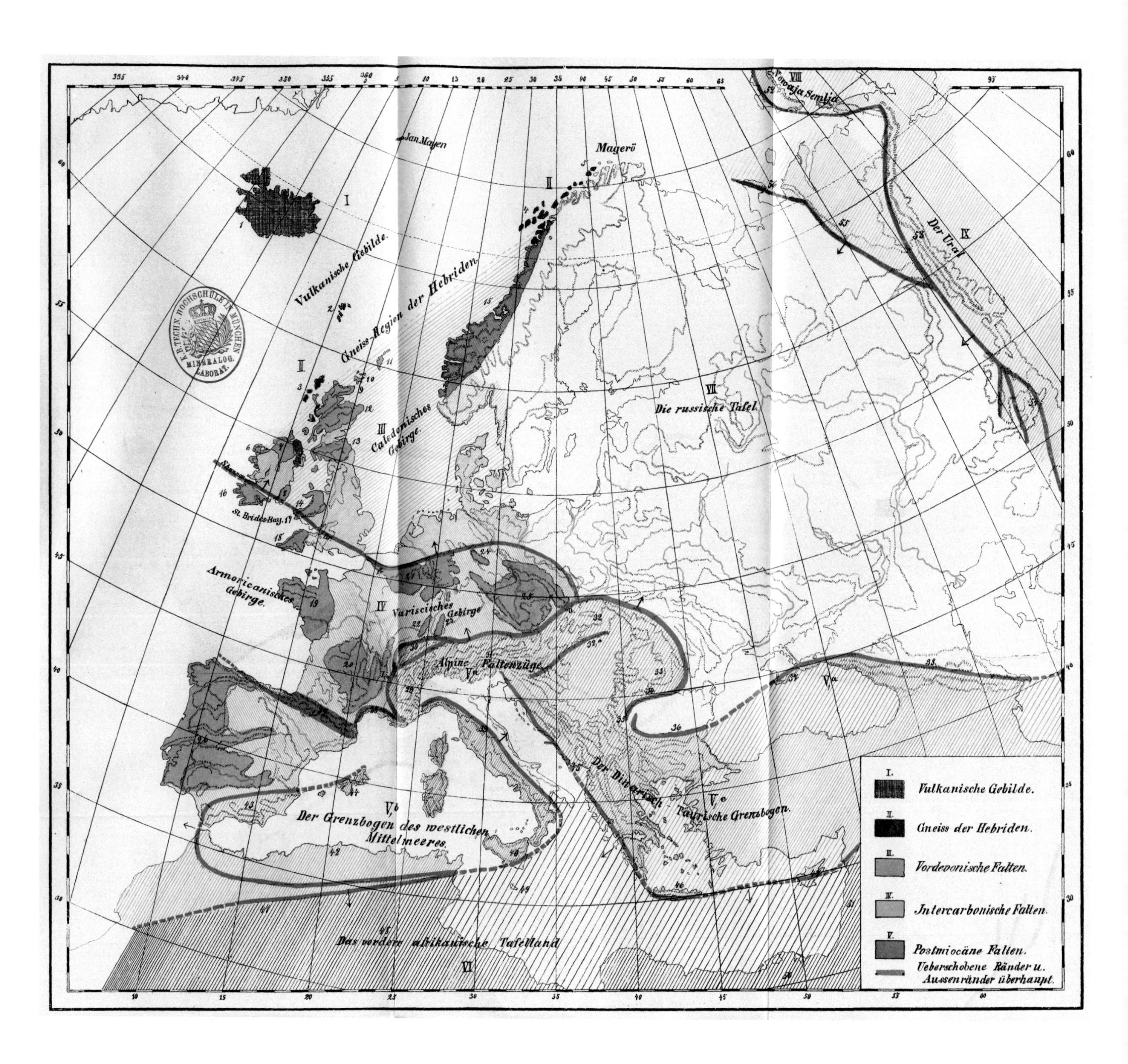

Building a continent. Three great mountain systems that make up Western and Central Europe: Caledonide Mountains in the north (green), formed before the Devonian (i.e. more than 416 million years old); Armorican and Variscan arcs in the middle (orange), built before the Permian (i.e. 250 million years ago); and the Alpides in the south (purple). Drafted by Suess, 1893.

Über unterbrochene Gebirgsfaltung, Sitzungsberichte der Kaiserlichen Akademie der Wissenschaften, Mathematisch-Naturwissenschaftliche Classe, I. Abtheilung, Wien, 94, 1886, pp. 116 – 117

Es sind drei hauptsächliche Zonen der Faltung in Mittel-Europa vorhanden. Die erste bildet das caledonische Gebirge und ist von vordevonischem Alter. Durch die zweite wurde das armoricanische Gebirge im Westen und das variscische Gebirge im Osten in vorpermischer Zeit aufgebaut; sie sind eingestürzt und die Horste sind abradirt, doch ist auch jüngere Senkung und nachträgliche, jüngere Faltung an vielen Orten sichtbar. Die dritte Zone sind die Pyrenäen und die Alpen. Auch die Alpen besitzen bereits Einbruchsfelder; die Senkung von Wien ist ein Beispiel. Stets ist die faltende Kraft nordwärts gerichtet gewesen, seit der Aufrichtung des caledonischen Gebirges bis zu den jüngsten Faltungen in den Alpen, und die wiederholten Einbrüche so wie die wiederholte Anlage neuer Falten haben hierin in dem betrachteten Gebiete keine Aenderung herbeigeführt.

In Middle Europe three main zones of folding exist. The first one are the Caledonide mountains, of pre-Devonian age. The second are of pre-Permian age and formed the Armorican mountains in the West and the Variscan mountains in the East. They have collapsed and their remains been smoothed, but earlier subsidence and additional younger folding can be found in many different places. The third zone includes the Pyrenees and the Alps. Subsidence areas are also found in the Alps; one example is the subsidence of Vienna. The folding direction has always been to the North, from the lifting of the Caledonide mountains to the youngest folding of the Alps, and neither the repeated subsidences nor the repeated addition of new folds has caused any changes in the observed area.

Caledonide Mountains. The major Caledonian Unconformity at Siccar Point, Scotland, U.K., where vertical beds of Silurian greywacke (425 million years old) are overlain by gently sloping strata of Devonian Old Red Sandstone (345 million years old).

Ueber neuere Ziele der Geologie,
Abhandlungen der Naturforschenden Gesellschaft zu Görlitz, 20, 1893, p. 11

Während alle die früher genannten grossen Grenzbogen Eurasiens nach Süden gefaltet sind und auch durch ganz Central-Asien die südliche Faltungsrichtung vorherrscht, sieht man im mittleren Europa die Faltenzüge gegen Norden gerichtet, und zwar ist Europa zu wiederholten Malen gefaltet worden und immer wieder nach Norden. Beginnen wir im Nordwesten. Island ist wie Jan Mayen von vulkanischer Natur. Die westlichen Hebriden, ganz Norwegen, die Lofoten bis Magerö und bis zum Nordcap hinauf bestehen aus uraltem Gneiss. Tritt man jedoch von den Inseln auf das nordwestliche Schottland herüber, so begegnet man sofort Gesteinslagen, welche völlig überstürzt sind, in verkehrter Lagerung auf das alte Gneissgebiet hinaufgeschoben wurden und welche uns den Aussenrand eines grossen Faltenzuges darstellen. Dieser Faltenzug streicht im Allgemeinen gegen Nordost. Er umfasst einen grossen Theil von Irland, Wales und Theile von England, ganz Schottland und findet seine Fortsetzung in den westlichen Faltenzügen von Norwegen. Es war dies einmal ein einheitlich gefaltetes Hochgebirge, von welchem wir heute nur mehr Trümmer zu erkennen im Stande sind, und zwischen diesen Trümmern fluthet heute das Meer. Wir nennen es das caledonische Gebirge.

While all large mountain ranges at the Eurasian rim mentioned earlier are folded towards the South and in Central Asia the southern folding direction also dominates, the folded ranges in Middle Europe are directed to the North. In fact, Europe has been folded repeatedly and each time towards the North. Let us start in the northwest. Iceland is volcanic, as is Jan Mayen Island. The western Hebrides, all of Norway, Lofoten up to Magerö and to the North Cape consist of age-old gneiss. However, as soon as one steps from the islands over to north-western Scotland, one is immediately confronted with layers of rock that are completely inverted, pushed onto the old gneiss area in reverse layering, which show us the outer rim of a huge fold system. This fold system generally points north-eastward. It includes a large part of Ireland, Wales and parts of England, all of Scotland, and continues into the western fold systems of Norway. At some stage this was a uniformly folded high mountain range, of which we can identify only some remains in the present day, and between the remains the sea now spreads. We call it the Caledonide Mountains.

Über unterbrochene Gebirgsfaltung, Sitzungsberichte der Kaiserlichen Akademie der Wissenschaften, Mathematisch-Naturwissenschaftliche Classe, I. Abtheilung, Wien, 94, 1886, pp. 112–114

Dieses Gebirge ist jünger als die caledonischen Züge. Der grösste Theil der flötzführenden Carbonschichten hat an den Faltungen und Überschiebungen theilgenommen; die permischen Sedimente liegen flach. Es ist vielleicht von spät-carbonischem, jedenfalls von vorpermischem Alter. Von Frome bis Exeter ist es abgebrochen und die Fortsetzung gegen Ost ist unter jüngeren Ablagerungen begraben. Dem Horste von Cornwall und Devon entspricht aber gegen Süden noch ein anderer, im gleichen Sinne gefalteter Horst auf französischem Boden, welcher den Cotentin und die Bretagne sammt der Vendee umfasst. Bei Brest ist das Streichen W–O in dem übrigen Theile dieses Gebietes aber WNW–OSO, entsprechend den nördlichen Bogenstücken. [...] Wir nennen es das armoricanische Gebirge. [...] In der That sind die überschobenen Flötze, welche durch Belgien gegen Aachen streichen, als ein Stück des Aussenrandes eines zweiten Gebirgsbogens anzusehen, welcher seine hauptsächliche Faltung ebenfalls in spätcarbonischer, jedenfalls vorpermischer Zeit vollendet hat und ebenfalls später in Trümmer gebrochen worden ist. [...] Beide liegen im Lande der Varisker, und der grosse Bogen mag das variscische Gebirge heissen.

This mountain range is younger than the Caledonide system. Most of the seam-bearing Carboniferous layers participated in the foldings and overthrusts, the Permian sediments lay flat. It is possibly late-Carboniferous, certainly pre-Permian in age. From Frome to Exeter it is interrupted and the continuation eastwards is buried under younger deposits, but the geologic horst of Cornwall and Devon matches another horst further south, on French territory, which was folded in the same way and includes Cotentin, Brittany and the Vendée. At Brest, the W–E direction of this mountain range becomes WNW–ESE, matching the northern parts of the arc. [...] We call it the Armorican Mountains. [...] The overthrusted seams that run through Belgium towards Aachen must indeed be seen as the outer rim of a second mountain arc, which had also completed its main folding in the late-Carboniferous, certainly pre-Permian and was also later broken up. [...] Both lie in the land of the Varisci, and the large arc may be named the Variscan Mountains.

Armorican and Variscan arcs. Angular unconformity in La Marette quarry at Saint-Malon-sur-Mel, Brittany, France, where Neoproterozoic rocks (more than 550 million years old) are overlain by Ordovician rocks (circa 475 million years old).

Alpine foldbelt. Steeply inclined folds of Lower Cretaceous flysch sediments (circa 115 million years old) at the northern rim of the Alpine orogen, Dopplerhütte, Lower Austria.

Über neuere Ziele der Geologie,
Abhandlungen der Naturforschenden Gesellschaft zu Görlitz, 20, 1893, pp. 13 – 14

Wir schreiten weiter gegen Süden und erkennen nun leicht, dass die Alpen mit den Karpathen nichts Anderes sind, als ein drittes System ähnlicher Bögen. Mit überstürzten Aussenrändern ziehen die Falten der Alpen von der Durance her durch die Schweiz und Bayern und Oesterreich. Sie sind deutlich in ihrer Entwicklung nach Norden gehemmt durch die entgegenstehenden Horste, das ist durch die Bruchstücke des variscischen Bogens. So stauen sich die Falten der Alpen zunächst an dem abgebrochenen Ostrande des französischen Central-Plateau, der sich bei Lyon erhebt, dann an einer kleinen Gneissklippe, welche bei Dôle in der Gegend von Besancon heraufragt, hierauf an dem südlichen Rande des Schwarzwaldes und sobald sie das südliche Ende der böhmischen Masse umzogen haben, wenden sie sich, gleichsam frei geworden, in dem grossen karpathischen Bogen gegen Norden. Dieser Faltenzug heisst der Faltenzug der Alpen. Er ist weit jünger als die früheren und die Haupt-Epoche seiner Faltung liegt in der Tertiärzeit.

We continue further towards the South and can now easily recognize that the Alps and the Carpathians are nothing but a third system of similar arcs. With inverted outer edges, the folds of the Alps run from the Durance river through Switzerland, Bavaria and Austria. They are clearly limited in their development towards the North by the opposing horsts, which are pieces of the Variscan arc. The folds of the Alps at first accumulate at the broken-off eastern edge of the French central plateau, which rises at Lyon, then at a small cliff of gneiss, which towers at Dôle near Besançon, later at the southern edge of the Black Forest, and as soon as they have rounded the southern end of the Bohemian mass, they turn, quasi released, in the large Carpathian arc towards the North. This system is known as the fold system of the Alps. It is much younger than the earlier ones and its main folding stage dates back to the Tertiary.

Horsts. Road cut southwest of Burhaniye, near Edremit, Turkey. The Lower Miocene rocks (23 to 15 million years old) are cut and displaced by numerous normal faults. The wedge in the centre of the picture that narrows towards the top, flanked by two sets of dropped beds, is a small-scale version of a horst, an old miner's term, introduced into geology by Suess.

The Face of the Earth Vol. I., Clarendon Press, Oxford, 1904, p. 126

If the outer borders of two fields of subsidence approach each other so that a ridge is left between them, on both sides of which the two areas of depression descend more or less in the form of steps, then we have what we shall distinguish, making use again of a common mining word, as a *horst*, in this case a *horst of the first order*, as opposed to the subsidiary horsts which occur here and there between networks of fracture. As *horsts of the first order* we may mention for example the Schwarzwald, the Vosges, Morvan, and the Kaibab Plateau in Colorado.

Basins of subsidence. The western rim of the southern Vienna Basin, Lower Austria, a pull-apart basin, which formed when the Western Carpathians moved away from the Eastern Alps during the Alpine orogeny.

Der Boden der Stadt Wien, Braumüller, Wien, 1862, pp. 19 – 20

Da sich nun in den Alpen einerseits und in den Karpathen andererseits die einzelnen Gesteinszonen in ihrer Richtung, wie in ihrer Beschaffenheit so genau entsprechen, dürfen wir es mit Gewissheit aussprechen, dass beide Gebirgszüge einer und derselben geologischen Einheit angehören, wenn ich mich so ausdrücken darf, und dass sie, durch einerlei Erscheinungen gebildet, erst später von einander getrennt wurden. [...] Wien liegt nicht zwischen zwei selbständigen Gebirgszügen, sondern mitten in den Alpen selbst, zwischen der Centralkette und der Grauwackenzone einerseits und der Sandsteinzone andererseits, unmittelbar auf dem Gebiete der eingesunkenen Kalksteinzone.

Since the Alps, on the one hand, and the Carpathians, on the other, match so exactly in terms of direction and texture of their rock zones, we can with certainty pronounce both mountain ranges as belonging to the same geologic unit, and, if I may say so, formed by the same phenomena and not separated until later. [...] Vienna is not situated between two separate mountain ranges but rather in the middle of the Alps, between the central range and the greywacke zone, on one side, and the sandstone zone on the other, directly in the area of the sunken limestone zone.

T*his descent of parts of the earth's crust seems to be the true origin of the great oceanic basins.* Sometimes the contour of the sunken area follows the trend of a folded mountain chain; at another time it may cut right across it. In smaller examples the outline very often takes a more or less irregularly circular or elliptical form. The descent of a considerable area, forming a large row depression, demands a certain part of the existing volume of oceanic waters for the filling of the new depth. The consequence is the sinking of the oceanic surface all over the planet, and the *apparent step-like rising of coast lines.* Thus is explained the apparently episodic elevation of whole continents, without any disturbance of horizontality, or the least alteration of the net of watercourses spread over the land. It is in this sense alone that a certain balance of "elevation" and "subsidence" might be conceded.

Are Great Ocean Depths Permanent? Natural Science, 2, London, 1893, p. 180

Formation of marine basins. Temple of Athena in the city of Assos. The Straits of Müsellem are located between Lesbos and the Turkish coast and include the Graben of Edremit that forms part of the Aegean Sea, delimited by step faults.

Earthquakes and faults. Burning houses, ships and masses of debris are piled up amidst tsunami flood waters in Kisenuma city, Miyagi prefecture, Japan, 12 March 2011. An earthquake of magnitude 8.9 hit northern Japan on 11 March 2011.

Die Erdbeben Nieder-Österreichs, Denkschriften der Kaiserlichen Akademie der Wissenschaften, Mathematisch-Naturwissenschaftliche Classe, 33, 1873, p. 83

Die Benützung localer Berichte hat eine grosse Anzahl von Erdbeben erkennen lassen, welche bisher übersehen wurden, während andererseits Angaben, welche in den bisherigen Listen (so insbesondere auf Grund der Verzeichnisse von Cotte in den Annales de Physique) wiederholt aufgeführt werden, als ganz zweifelhaft angesehen werden müssen, weil die sonst genauen localen Überlieferungen ihrer gar nicht erwähnen. Bei vielen älteren Erdbeben lässt sich allerdings oft nicht erkennen, ob sie ihren Ursprung innerhalb dieses Gebietes hatten oder nur Undulationen waren, welche von entfernten Katastrophen herrührten, sobald aber genauere Angaben beginnen, nämlich vom Jahre 1267 an, findet man die selben Punkte genannt, welche auch heute von Erdbeben besonders häufig getroffen werden.

The use of local reports has revealed a great number of earthquakes, which have been overlooked so far, while indications which were repeated in previous lists (especially those based on Cotte's lists in the Annales de Physique), must be viewed as rather doubtful, since the otherwise precise local reports do not mention them. For many older earthquakes, however, it is often hard to determine if their origin was in the same area or if these were just undulations which resulted from distant catastrophes. When more detailed reports become available, namely from 1267 onwards, the same places are mentioned which today also experience earthquakes most often.

Horizontal movements. The Glarus overthrust at Martins Loch, canton of Glarus, Switzerland. The pyramidal peaks (290 to 270 million years old) have been overthrust onto a horizontal surface across a thin band of rocks (circa 160 million years old) and these in turn rest on rocks that are younger than 55 million years. This implies a minimum of 40 to 60 km northward displacement of the rocks on top.

Die Entstehung der Alpen, Braumüller, Wien, 1875, p. 25

Immer deutlicher zeigt sich schon bei diesen ersten Betrachtungen, dass gleichförmige Bewegungen grosser Massen im horizontalen Sinne einen viel wesentlicheren Einfluss auf die heutige Gestaltung des Alpensystems gehabt haben, als die bisher allzusehr betonten verticalen Bewegungen einzelner Theile, d. h. die unmittelbaren Erhebungen durch eine radial aus dem Inneren des Planeten auf seine Oberfläche wirkende Kraft.

From these initial observations alone it has become clearer that uniform horizontal movements of large masses have exerted a much more significant impact on the current form of the Alpine system than the hitherto overemphasized vertical movements of individual parts, i.e. the immediate lifts caused by a radial force acting on the surface of the planet from within.

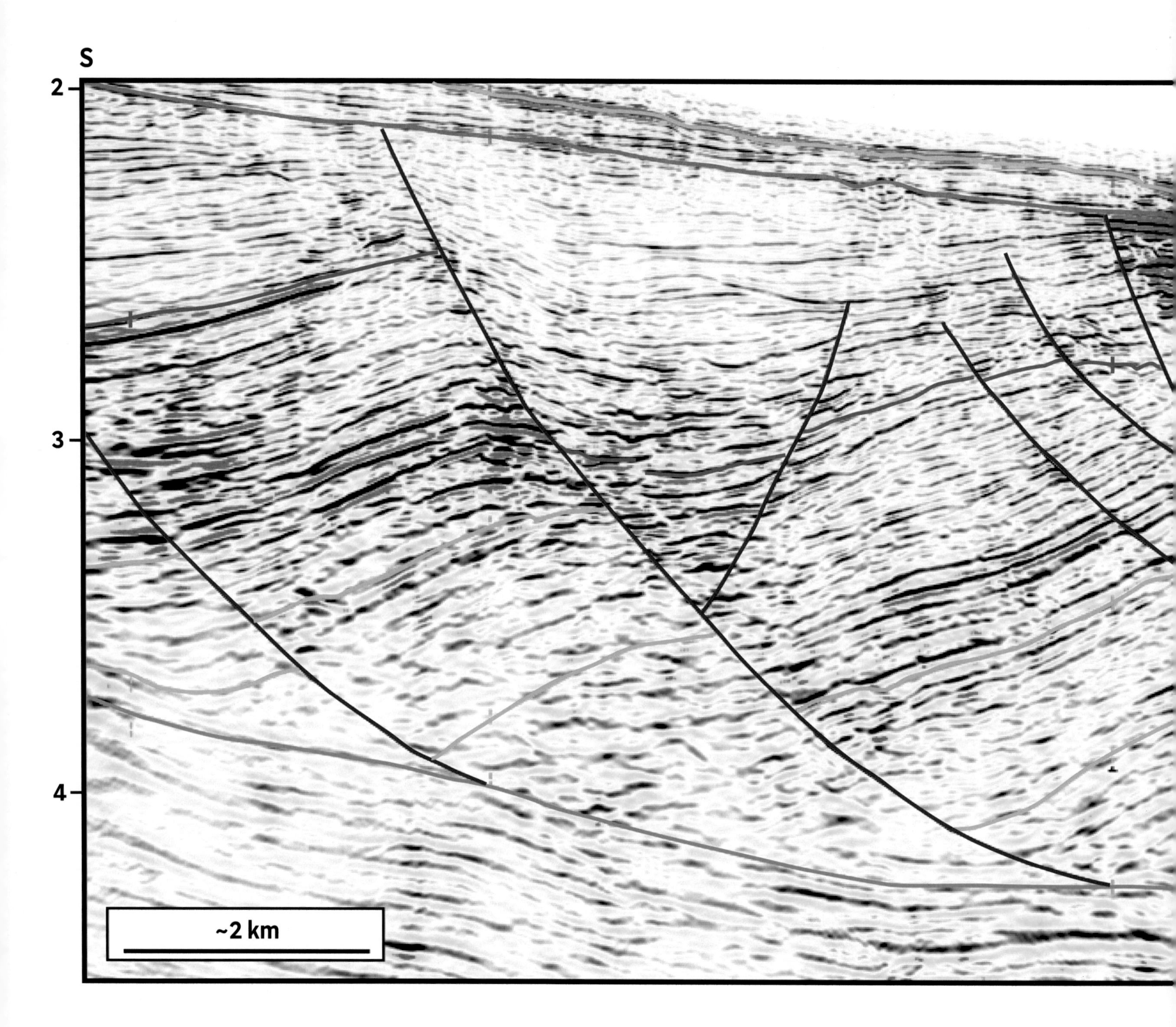

Listric faults. Seismic illustration of the central offshore segment of the Matruh Detachment system, Eastern Mediterranean. The curved dark blue lines indicate the listric faults.

The Face of the Earth Vol. IV., Clarendon Press, Oxford, 1909, p. 536

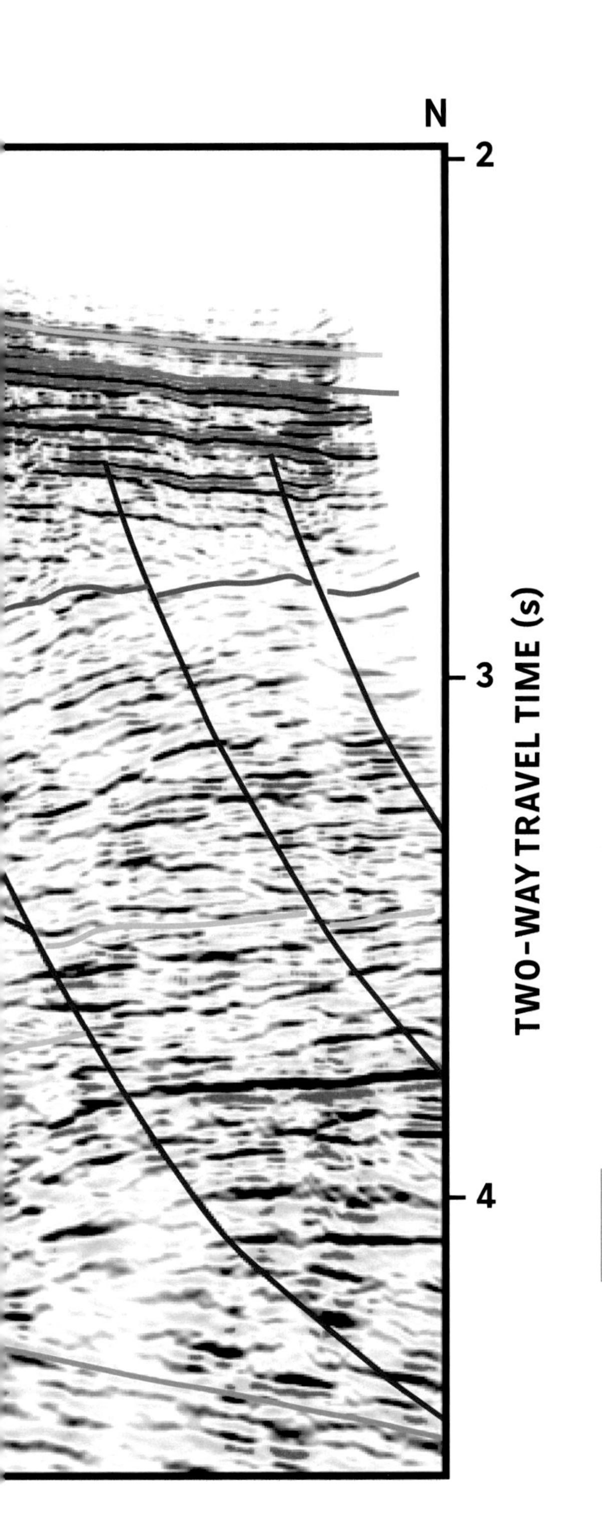

Planes of movement, such as the Failles du Carabinier, du Placard, and the others shown on fig. 44, cannot therefore be described as faults; neither are they produced by folding; they are planes of a special character. We will term them *listric planes* (λίστρον, a shovel).

Batholiths – the space problem. Granites of the Thaya Batholith (more than 550 million years old) at the eastern rim of the Bohemian Massif near Eggenburg, Lower Austria.

The Face of the Earth Vol. IV., Clarendon Press, Oxford, 1909, pp. 551 – 552

B*atholites*. This term was proposed (I, p. 168) for the great cake-like masses, composed chiefly of granite, which appear to be inserted in the stratified formations. [...], and now batholites may be described as intrusive masses, which are continued down into the 'eternal depths', in opposition to laccolites, which are lateral intrusions upon an alien foundation. Indeed, a consideration of the facts as they occur in nature leads to the certain conclusion that batholites have reached their present position (mise en place) by *melting and absorbing the adjacent rock*.

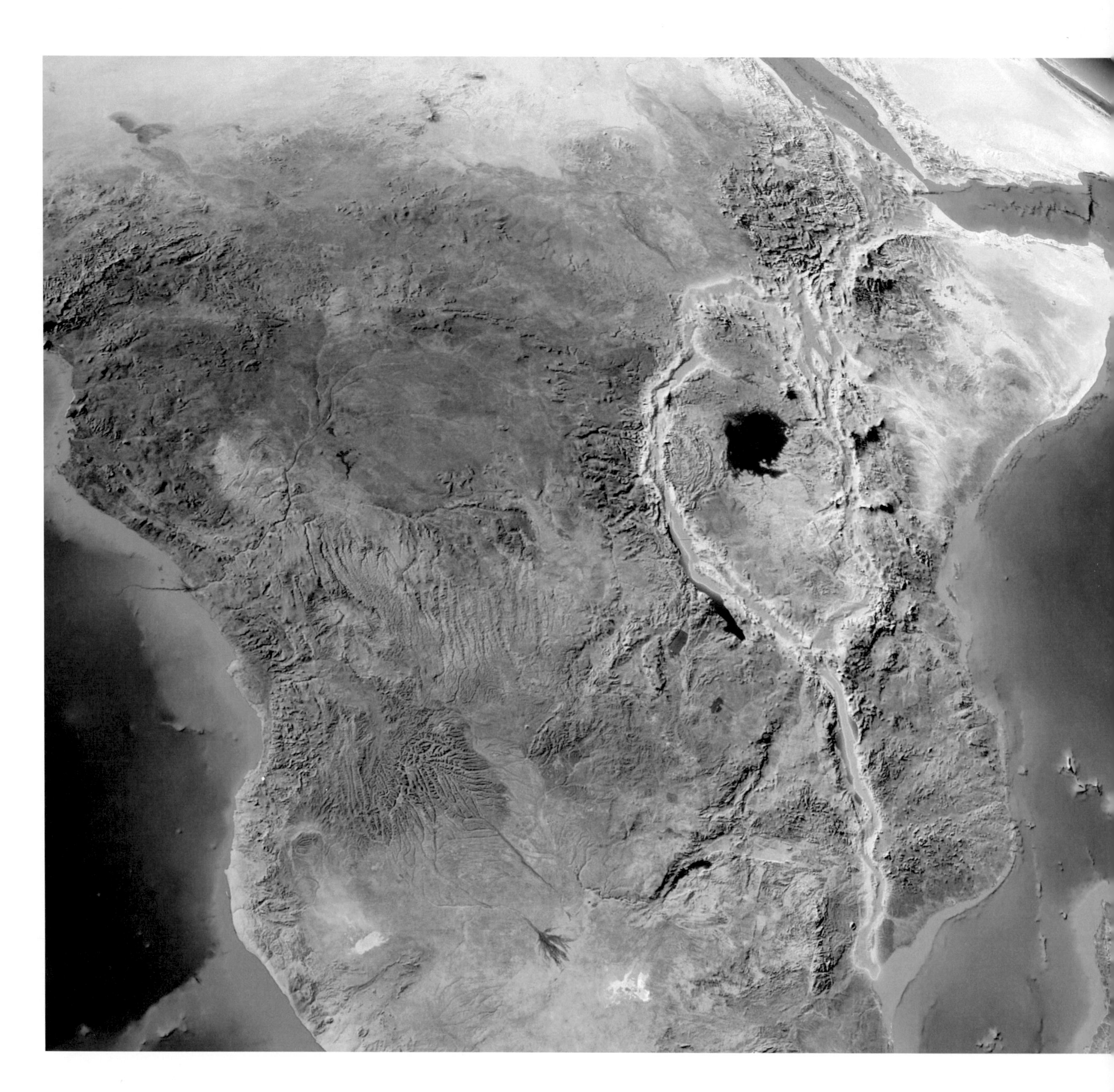

Continental breakup. The East African rift valley (highlighted in orange) indicating the initial stage of the breakup of Africa into two new tectonic plates, the Nubian Plate and the Somali Plate.

Beiträge zur geologischen Kenntniss des östlichen Afrika. Theil IV., Denkschriften der Kaiserlichen Akademie der Wissenschaften, Mathematisch-Naturwissenschaftliche Classe, Wien, 58, 1874, pp. 555, 577, 580

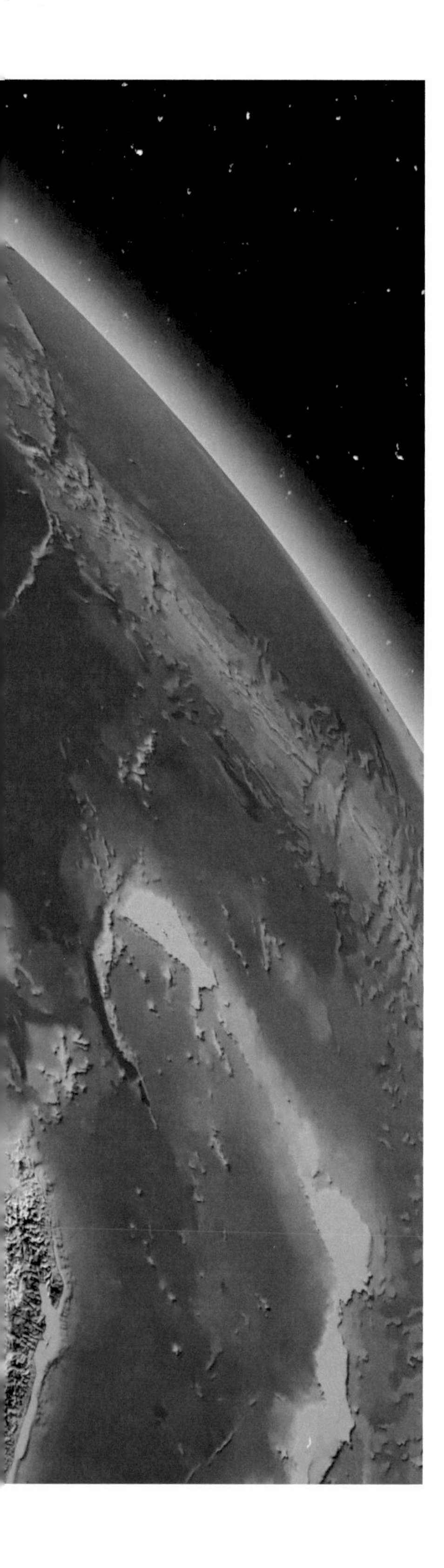

Die fortschreitende Erforschung des östlichen Afrika hat die Vermuthung wachgerufen, dass in diesem Theile des Planeten irgend eine ausserordentliche, nahezu im Meridian verlaufende Dislocation, eine Spalte oder ein Einbruch von sehr grosser Länge vorhanden sei. [...] Aber im Ganzen gleicht dieses Gebiet allem Anscheine nach mehr einer lange fortlaufenden Zone der Zertheilung der Erde in längliche Schollen und Trümmer, wie solche entstehen mag, wenn ein in grosser Tiefe vorhandener Spalt gegen oben in zahlreiche lange und sich maschenförmig durchkreuzende Klüfte wäre zersplittert worden, welche Trümmer und Schollen dann in ihrer Gesammtheit, aber zu ungleicher Tiefe abgesunken wären. [....] Auf der ganzen Linie aber, von Süd bis Nord war in jüngster Zeit und ist zum Theile noch heute die vulkanische und seismische Thätigkeit eine sehr beträchtliche. [...] So weit mir die Sachlage aus den vorliegenden Darstellungen bekannt ist, scheint es wohl, als sei die Bildung so grosser Spalten nur erklärbar durch das Vorhandensein einer Spannung, deren Richtung senkrecht steht auf der Richtung der Spalte und welche Spannung in dem Augenblicke des Berstens, d. i. des Aufreissens der Spalte ihre Auslösung findet.

The progressive exploration of eastern Africa has given rise to the idea that in this part of the globe some kind of extraordinary dislocation, a rift or subsidence of great length, running almost along the meridian, must be present. [...] Yet, overall, this area rather seems to resemble a long, continuous zone where the earth has been fragmented into oblong slabs and fragments of rock, as might form when a very deep rift is split upwards into numerous criss-crossing fissures, and the fragments and slabs in their entirety would then have subsided, but to uneven depths. [...] Along the entire line, from south to north, there has been considerable volcanic and seismic activity in recent times and some continues today. [...] As far as I understand the situation from the available reports, it appears that the emergence of rifts of such a size can only be explained by the presence of a tension that is vertical to the direction of the rift and which finds its release when the rift bursts and ruptures.

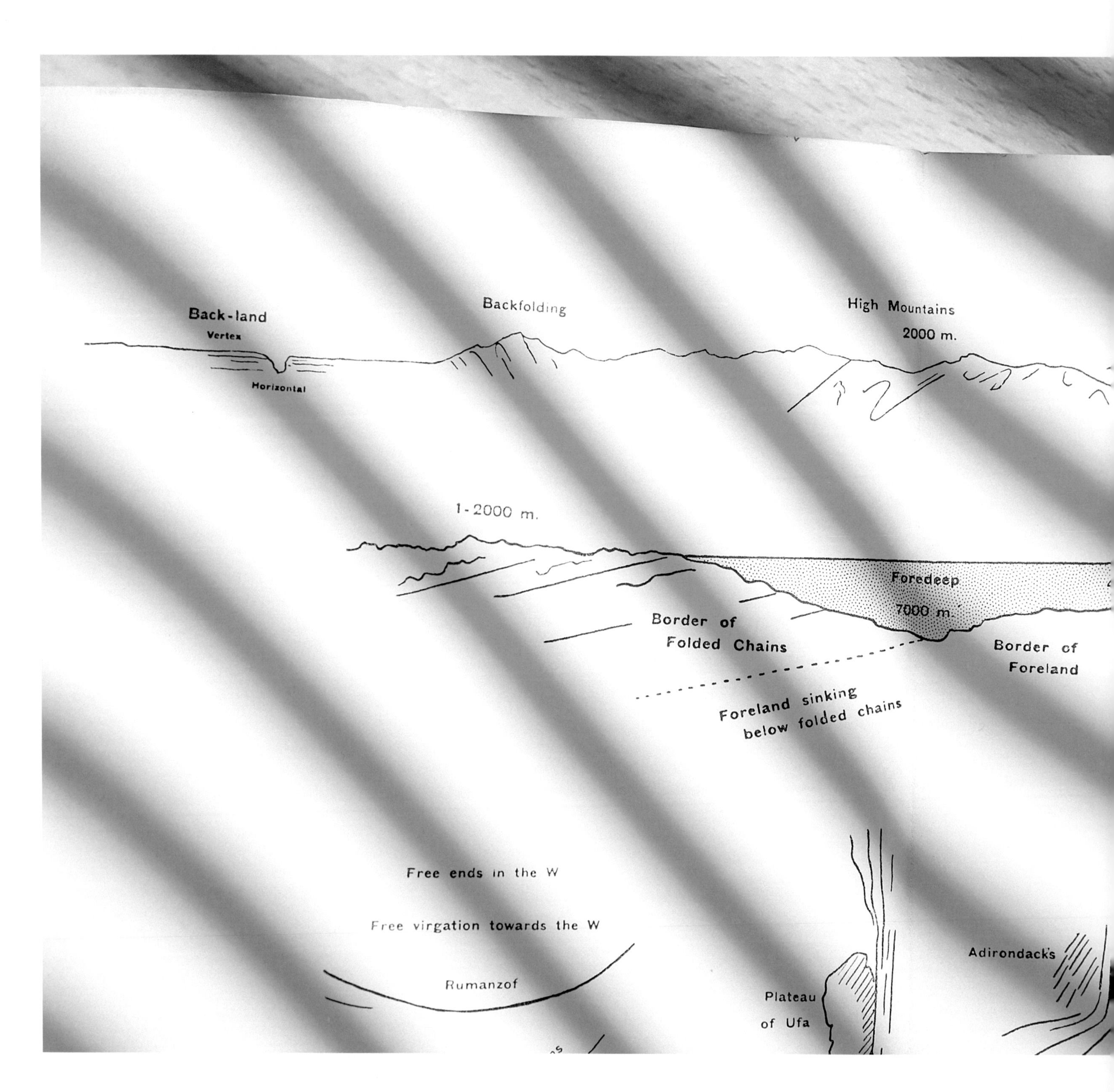

Foredeeps. Original sketch by Suess of foredeep sinking, anticipating the concept of plate tectonics.

The Face of the Earth Vol. IV., Clarendon Press, Oxford, 1909, p. 506

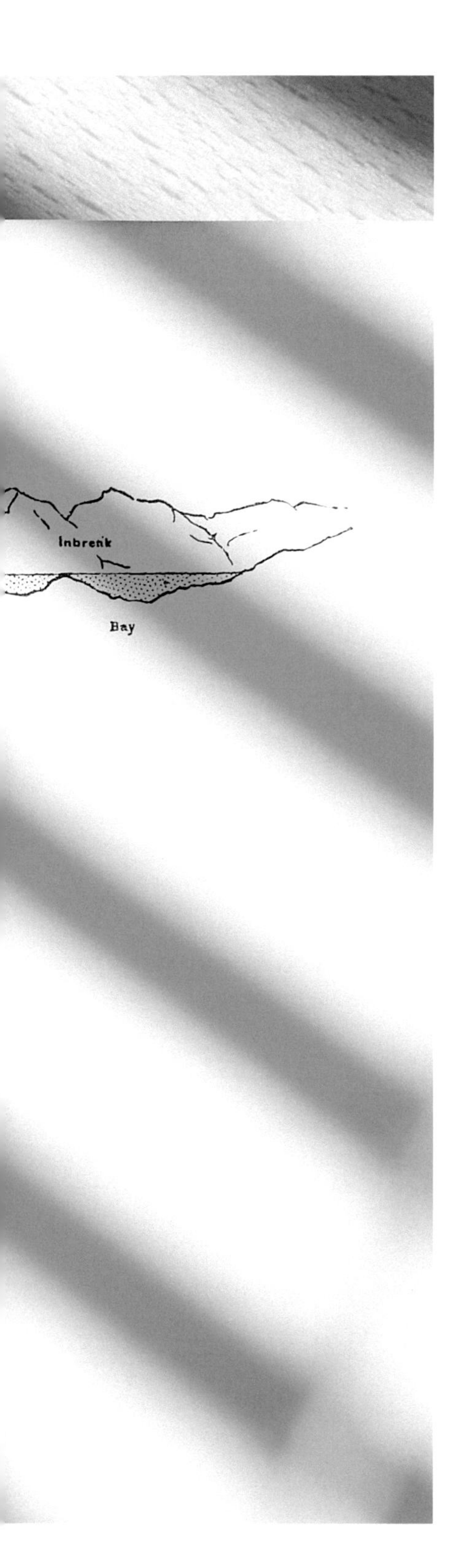

T*he outer border of the foredeep is the arcuate border of a subsidence of the lithosphere, and the inner border of the foredeep is the outer border of the folded range* which has advanced from the direction of the land over this deep.

Layers of the earth. Atmosphere, hydrosphere, biosphere and lithosphere at Capetown, South Africa.

Die Entstehung der Alpen, Braumüller, Wien, 1875, pp. 158–159

Die Hydrosphäre gibt Dünste in die Atmosphäre ab, diese verdichten sich und kehren zurück. Die porösen Theile der Lithosphäre nehmen Wasser auf lassen es circuliren und als Quellen wieder aufsteigen. Viel Wasser wird chemisch gebunden. Fortwährend werden lose Theile der Lithosphäre an tieferen Stellen getragen und aus der Wechselwirkung der Hydrosphäre und der Lithosphäre ist neue, wenn auch unvollständige Sphäre der von geschichteten Felsarten als die oberste Sphäre der Lithosphäre entstanden. [...] Eines scheint fremdartig auf diesem grossen, aus Sphären gebildeten Himmelskörper, nämlich das organische Leben. Aber auch dieses ist auf eine bestimmte Zone beschränkt, auf die Oberfläche der Lithosphäre. Die Pflanze, welche ihre Wurzeln Nahrung suchend in den Boden senkt und gleichzeitig sich athmend in die Luft erhebt, ist ein gutes Bild der Stellung organischen Lebens in der Region der Wechselwirkung der oberen Sphären und der Lithosphäre, und es lässt sich auf der Oberfläche des Festen eine selbständige Biosphäre unterscheiden.

The hydrosphere releases vapours into the atmosphere that condense and return as water. The porous parts of the lithosphere absorb water, let it circulate and release it as springs. Most of the water is chemically bound. Loose parts of the lithosphere are constantly carried downwards, and from the interaction of the hydrosphere with the lithosphere a new, if incomplete, sphere is formed by layered types of rock in the topmost sphere of the lithosphere. [...] One phenomenon seems strange on this large orb of many spheres: organic life. It is, however, also restricted to a particular zone, the surface of the lithosphere. The plant, which lowers its roots into the ground in the search for nutrients and at the same time grows, breathing, into the air, is a good image of the position of organic life in the area where the upper spheres and the lithosphere interact, and we can distinguish an independent biosphere on the firm surface.

Further, we assume the existence of three zones or envelopes as determining the structure of the earth, namely, the barysphere or the Nife (Ni-Fe), Sima (Si-Mg), and Sal (Si-Al). This division differs from the classification which has been proposed by distinguished American petrographers, in the separation of the metallic barysphere (Nife).

The Face of the Earth Vol. IV., Clarendon Press, Oxford, 1909, p. 544

The interior of the planet. Widmanstätten pattern on the etched and polished surface of the Hraschina iron meteorite, Natural History Museum Vienna. Iron meteorites were used by Suess to infer the composition of the earth's core.

Erinnerungen, S. Hirzel, Leipzig, 1916, p. 354

So knapp griffen zu gewissen Zeiten die Aufgaben meiner wissenschaftlichen und beruflichen und meiner öffentlichen Tätigkeit ineinander, und nur durch die strengste Abgrenzung war es möglich, im eigenen Denkvermögen Mengungen zu vermeiden. Nie durch die 45 Jahre meiner Lehrtätigkeit erinnere ich mich, zu meinen Hörern ein Wort über Tagespolitik gesprochen zu haben. Hier muss ich ein Wort über meine Wählerschaft sagen. Vom Beginne meiner öffentlichen Tätigkeit im Jahre 1862 bis zu ihrem Ende im Jahre 1897 hat meines Erinnerns nie ein Wähler gewagt, mich um die Befürwortung persönlicher Interessen bei der Regierung anzugehen – mit einer einzigen Ausnahme, in der ich glatt ablehnen konnte.

At certain times the tasks of my scientific and professional activities were closely interlinked with those of my public service, and it took the most rigorous delimitation to avoid mingled thinking. Never in the 45 years of teaching do I recall having spoken to my students about the politics of the day. Here I must say a word about my voters. From the beginning of my public service in 1862 to its end in 1897, no voter, as far as I can remember, ever dared to approach me to ask for assistance with his personal interests vis-à-vis the government – with one exception, which I was able to refuse outright.

Public affairs. The Colonnaded Hall at the Austrian Parliament where Suess was a member of the former Reichstag from 1873 to 1897. The monolithic Corinthian pillars, each weighing around 16 tons, are made of Triassic Adnet marble from Salzburg, Austria.

The Face of the Earth Vol. I., Clarendon Press, Oxford, 1904, p. 604

The breaking up of the terrestrial globe, this it is we witness. It doubtless began a long time ago, and the brevity of human life enables us to contemplate it without dismay.

Volcanic eruption, Fimmvörðuháls, southern Iceland.

Back cover | Blasting CME: This LASCO C2 image, taken on 8 January 2002, shows a widely spreading coronal mass ejection (CME) as it blasts more than a billion tons of matter out into space at millions of kilometres per hour. The C2 image was turned 90 degrees so that the blast seems to be pointing down. For effect, an EIT 304 Angstrom image from a different day was enlarged and superimposed on the C2 image so that it filled the occulting disk.

Idea Thomas Hofmann
Concept Thomas Hofmann, Günter Blöschl, Lois Lammerhuber, Werner E. Piller, A. M. Celâl Şengör

Quotes Eduard Suess
Quotes edited by A. M. Celâl Şengör

Authors Thomas Hofmann, Günter Blöschl, Werner E. Piller, A. M. Celâl Şengör
Translation Brigitte Scott
Proofreading Gemma Carr
Review Jonathan Bamber, Bruno Merz, Fabrizio Storti

Photography Lois Lammerhuber
Illustrations and images on pages 8–9 Sean R. Heavey **10** © eoVision/DigitalGlobe, 2014, distributed by e-GEOS **18–19** Archive of the Department of Geodynamics and Sedimentology (University of Vienna) **24–25, 86–87** Kurt Stuewe and Ruedi Homberger **26** USGS/EROS; http://visibleearth.nasa.gov **34** Courtesy of RZB **48** ©UCAR, image courtesy of Matthias Rempel, NCAR **50, 92–93** © eoVision/ NASA / GSFC, 2014 **56** Georg Riha **64** © eoVision/ Image Science & Analysis Laboratory, NASA Johnson Space Center, 2014 **66–67** Monika Brüggemann-Ledolter & Gerhard W. Mandl (Geological Survey of Austria) **70** Abhandlungen der Naturforschenden Gesellschaft zu Görlitz, 20, 1893 **84–85** Str/epa/picturedesk.com **88–89** OMV **94–95** The Face of the Earth, Volume 5, 1924 **102** Skarphedinn Thrainsson, www.skarpi.is **Back cover** Courtesy of SOHO/LASCO consortium. SOHO is a project of international cooperation between ESA and NASA.

Art director Lois Lammerhuber
Graphic design Martin Ackerl, Lois Lammerhuber
Typeface Lammerhuber (beta) by Titus Nemeth
Digital post production Birgit Hofbauer
Project coordination Johanna Reithmayer

Printing and Binding Gorenjski tisk storitve, Kranj, Slovenia
Paper Hello Fat matt, 170 g/m²

Responsible at EUROPEAN GEOSCIENCES UNION Günter Blöschl
EUROPEAN GEOSCIENCES UNION Luisenstraße 37, 80333 Munich, Germany
www.egu.eu

Managing director EDITION LAMMERHUBER Silvia Lammerhuber
EDITION LAMMERHUBER Dumbagasse 9, 2500 Baden, Austria
edition.lammerhuber.at

THE FACE OF THE EARTH – Process and Form – is the theme of the 2014 General Assembly of the European Geosciences Union.

Geological Survey of Austria

The authors would like to thank Christian Cermak, Lorenz Dobramysl, Bruno Granier, Valentin Klausburg, Richard Lein, Nalan Lom, Oya Şengör, Philipp Strauss, Kurt Stüwe, Gabor Tari.

The authors would also like to thank the following institutions for financial support: European Geosciences Union, Austrian Academy of Sciences, Copernicus Gesellschaft e.V., Vienna University of Technology and Federal Ministry of Science, Research and Economy.

Copyright 2014 by EDITION LAMMERHUBER ISBN 978-3-901753-69-5
All rights reserved. This work and any images therein may not be reproduced in whole or in part. Any exploitation is only permitted with the publisher's and authors' written consent.

Thomas Hofmann is Head of the library, archive and press of the Geological Survey of Austria. He has published numerous articles on geotopes as well as on various aspects of the history of the earth sciences, with a special focus on themes related to the Geological Survey of Austria. He is an active networker in promoting geosciences to a wider audience. His regular writings include many popular science articles, plus non-fiction books, travel guides and other specialist literature.

Günter Blöschl is Professor of Hydrology, Director of the Centre for Water Resource Systems, and Head of the Institute of Hydraulic Engineering and Water Resources Management at the Vienna University of Technology. He has published extensively on subjects related to hydrology and water resources. He is Fellow of the American Geophysical Union, member of the German Academy of Science and Engineering, and recipient of an Advanced Grant of the European Research Council (ERC). Currently, he is the President of the European Geosciences Union.

Lois Lammerhuber is an Austrian publisher and a self-taught artist. Since 1984 he has worked closely with GEO magazine. He has published 2400 photo essays and 92 books which have garnered numerous prizes and other honours, including three Graphis Phototography Awards for world best reportage of the year. In 2014 he received the Republic of Austria's Decoration for Science and Art. Lois Lammerhuber has also authored countless radio broadcasts in Germany and Austria. He is a member of the Art Directors Club New York.

Werner E. Piller is Professor of Palaeontology and Historical Geology at the University of Graz. His research spans fields ranging from taxonomy, palaeoecology and palaeoceanography, biogeography, stratigraphy to molecular biology. He has published numerous scientific papers, books and journal issues. He is a full member of the Austrian Academy of Sciences, Chair of the Austrian National Committee of Geosciences and the Austrian National Committee of Geo/Hydro-Sciences, and President of the Regional Committee of Mediterranean Neogene Stratigraphy.

A. M. Celal Şengör is a geologist with interest in global tectonics. His interest in geology was awakened by Jules Verne and in tectonics by Eduard Suess (whom he read as a high school student). He studied under John F. Dewey and Kevin Burke in Albany, New York. His subsequent work was mainly on Asia with occasional forays into different continents including Europe, North America and Africa and theoretical tectonics. He has been a faculty member of Istanbul Technical University since 1981.

All publications of Eduard Suess, as well as those of the European Geosciences Union (EGU) are available online for free from the library of the Geological Survey of Austria.
http://opac.geologie.ac.at